Student Solutions Manual for HECHT'S PHYSICS: Calculus

SECOND EDITION

JERRY SHI
Pasadena Community College

EUGENE HECHT
Adelphi University

Australia • Canada • Mexico • Singapore
Spain • United Kingdom • United States

Assistant Editor: *Melissa Duge Henderson*
Marketing Manager: *Steve Catalano*
Marketing Assistant: *Christina de Veto*
Editorial Assistant: *Dena Dowsett-Jones*
Production Coordinator: *Dorothy Bell*

Cover Design: *Vernon T. Boes*
Cover Photo: *Joseph Drivas, The Image Bank*
Print Buyer: *Micky Lawler*
Printing and Binding: *Webcom Limited*

COPYRIGHT © 2000 by Brooks/Cole
A division of Thomson Learning
The Thomson Learning logo is a trademark used herein under license.

For more information about this or any other Brooks/Cole product, contact:
BROOKS/COLE
511 Forest Lodge Road
Pacific Grove, CA 93950 USA
www.brookscole.com
1-800-423-0563 (Thomson Learning Academic Resource Center)

All rights reserved. No part of this work may be reproduced, transcribed or used in any form or by any means—graphic, electronic, or mechanical, including photocopying, recording, taping, Web distribution, or information storage and/or retrieval systems—without the prior written permission of the publisher.

For permission to use material from this work, contact us by
Web: www.thomsonrights.com
fax: 1-800-730-2215
phone: 1-800-730-2214

Printed in Canada

10 9 8 7 6 5 4 3 2 1

ISBN 0-534-37248-1

Preface

This **Student Solutions Manual** was written to complement *PHYSICS: Calculus*, second edition, by Eugene Hecht. It is designed to assist you in working independently to master the material discussed in the text.

Every chapter in the main text contains approximately twenty Discussion Questions. A selection of the odd-numbered questions are listed with italic numerals, and the answers to those questions appear in this manual. The Discussion Questions are meant to help you explore and develop your conceptual comprehension of the material. If an answer is not apparent, reread the appropriate portion of the text. The answers provided in this manual should be consulted only as a last resort or, alternatively, as a check.

Every chapter in the main text also contains approximately twenty Multiple Choice Questions. These are designed to test and develop your ability to deal quickly and accurately with conceptual issues that require little or no mathematical processing. Read the questions carefully. Make no assumptions about what the author might have had in mind. Respond only to the question as stated. This manual contains answers to the odd-numbered Multiple Choice Questions which are also answered in the main text.

Every chapter in the main text also contains about eighty numerical problems grouped by specific topics. A representative selection of the odd-numbered problems have italic numerals. Complete solutions for all of these problems appear in this manual. Again you are advised to attempt to solve each problem before consulting the Student Solutions Manual. The Answer Section of the main text contains a selection of skeletal solutions. Study the solutions for similar problems before turning to the complete solutions provided in this manual.

This project would not have been possible without the great support and guidance of Beth Wilbur, Physics Editor at Brooks/Cole Publishing Company. Special thanks also go to Melissa Henderson, Ancillary Editor at Brooks/Cole; Kurt Norlin, of Laurel Technical Services; and Lorraine Burke, of HRService Group. Last but not the least, we wish to acknowledge Yiping Lu and Xidong Zhang, for skillfully typesetting part of this manual in AMS TEX.

Suggestions, corrections, and criticisms are greatly appreciated and should be sent to:
Professor Gene Hecht, Department of Physics, Adelphi University, Garden City, NY 11530; or Professor Jerry Shi, Physical Sciences Division, Pasadena City College, Pasadena, CA 91106. Good Luck!

Contents

1 An Introduction to Physics

Answers to Selected Discussion Questions

•1.1

Because on rare occasions there have been scientists who have simply lied about their work. The biologist who, in his laboratory, colored the butt of a white mouse with a black magic marker and then claimed to have performed a skin graft wasn't doing science. There are also a few cases of honest self-delusion in which what scientists have done has been utter nonsense and certainly not science.

•1.3

Laws transcend experience because they generalize. We know that something has happened the same way for years and from that we conclude that it will continue to happen that way. Laws are timeless, and that goes beyond experience. The same can be said about the application of law everywhere — it would be strange if certain physical laws only worked in Chicago. Still, we certainly don't test any of them everywhere.

•1.7

Data are the uninterpreted perceptions (as much as that's possible). Facts are the product of interpretation within the context of some world view, some theoretical understanding. "I see a disc of light there" becomes "I see the sun god in heaven" or "I see a star in space." These are statements of fact based on different world views. Since world views can be biased, facts can be wrong.

•1.9

The ultimate testing ground in physics is nature itself. If the universe is found experimentally to match the predictions of the theory, then it was "crazy enough."

Answers to Odd-Numbered Multiple Choice Questions

1. c **3.** d **5.** a **7.** a **9.** b **11.** a **13.** b
15. e **17.** c **19.** c **21.** b

Solutions to Selected Problems

1.1

Since 1 billion is equal to 1 000 000 000, or 10^9; 10 billion $= 10\,000\,000\,000 = 10 \times 10^9 = 10^{10}$.

1.11

Since $1\,\text{Å} = 0.1\,\text{nm}$, $5 \times 10^3\,\text{Å} = (5 \times 10^3\,\text{Å})(0.1\,\text{nm/Å}) = 5 \times 10^2\,\text{nm}$.

1.22

First, convert in.^{2} to cm^2: $1\,\text{in.}^2 = (2.54\,\text{cm})^2 = 6.45\,\text{cm}^2$. Thus the active odor-detecting area in the human nose is $(3/4\,\text{in.}^2)(6.45\,\text{cm}^2/\text{in.}^2) = 4.84\,\text{cm}^2$, which means that the dog's sensory area is $65\,\text{cm}^2/4.84\,\text{cm}^2 = 13$ times greater.

1.27

Since $1.000\,\text{in.} = 2.540\,\text{cm}$, $1.000\,\text{cm} = (1.000/2.540)\,\text{in.} = 0.393\,7\,\text{in.}$ Thus the conversion factor you need to multiply to convert cm to in. is $0.393\,7\,\text{in./cm}$. For example, $12.00\,\text{cm} = (12.00\,\text{cm})(0.393\,7\,\text{in./cm}) = 4.724\,\text{in.}$

1.37

The number of hours in a year is given by $(1\,\text{y})(365.25\,\text{d/y})(24\,\text{h/d}) = 8766\,\text{h}$. Thus the total number of particles shed in a year is $N = (6\times10^5\,\text{particles/h})(8766\,\text{h}) = 5.259\,6\times10^9$ particles. Since the total mass of these particles is $M = (1.5\,\text{lb})(0.453\,6\,\text{kg/lb}) = 0.680\,4\,\text{kg}$, the mass of each particle is

$$m = \frac{M}{N} = \frac{0.680\,4\,\text{kg}}{5.259\,6\times10^9\,\text{particles}} = 1\times10^{-10}\,\text{kg/particle} = 1\times10^{-7}\,\text{g/particle}\,.$$

Note that we retained only one significant figure in m, as the number of particles shed per hour (6×10^5 particles/h) is given in one significant figure only. The total mass of skin shed in 50 years will be $(50\,\text{y})(0.680\,4\,\text{kg/y}) = 0.3\times10^2\,\text{kg}$.

Note that we retained only one significant figure in m, as the number of particles shed per hour (6×10^5 particles/h) is given in one significant figure only. The total mass of skin shed in 50 years will be $(50\,\text{y})(0.680\,4\,\text{kg/y}) = 0.3\times10^2\,\text{kg}$.

1.42

The only way to display a quantity that large in whatever desired number of significant figures is to use scientific notation. The exact speed of light can be written as $2.997\,924\,58\times10^8$ m/s (with nine significant figures), which becomes 3×10^8 m/s with one significant figure, 3.00×10^8 m/s with three significant figures, 2.998×10^8 m/s with four significant figures, and $2.997\,924\,6\times10^8$ m/s with eight significant figures.

1.57

First convert the unit of each quantity to kg: $1.00\,\text{g} = (1.00\,\text{g})(10^{-3}\,\text{kg/g}) = 0.001\,00\,\text{kg}$, $1.00\,\text{mg} = (1.00\,\text{mg})(10^{-6}\,\text{kg/mg}) = 0.000\,001\,\text{kg}$, and $1.00\,\mu\text{g} = (1.00\,\mu\text{g})(10^{-9}\,\text{kg}/\mu\text{g}) = 0.000\,000\,001\,\text{kg}$. Now add these quantities up to obtain

$$0.001\,00\,\text{kg} + 0.000\,001\,\text{kg} + 1.00\,\text{kg} + 0.000\,000\,001\,\text{kg} = 1.001\,001\,001\,\text{kg}\,.$$

Since the least number of decimal places is two (in the quantity 1.00 kg), we can keep just two figures after the decimal point. Thus the answer should be 1.00 kg.

1.65

The plot is shown in the next page. Note that, as x crosses 2, $y(x)$ changes abruptly from $-\infty$ to $+\infty$, and so the curve is discontinuous at $x = 2$.

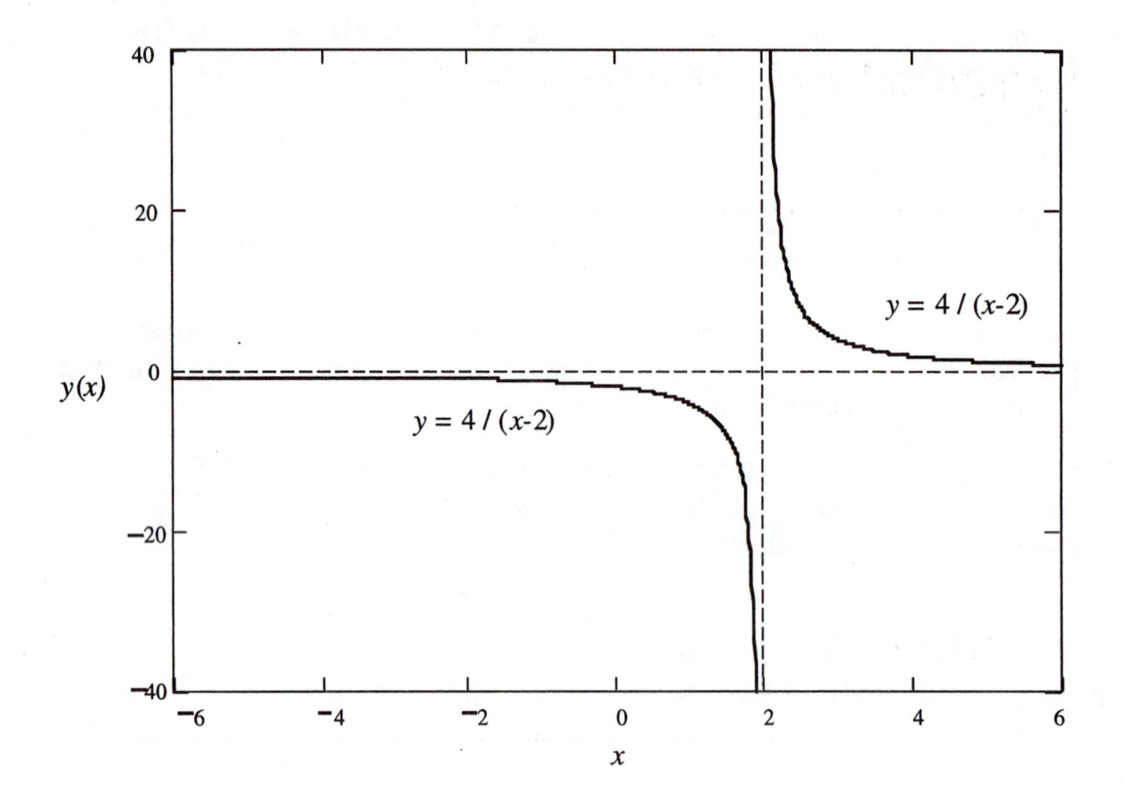
40
20
0
−20
−40
y(x)
y = 4 / (x-2)
y = 4 / (x-2)
−6
−4
−2
0
2
4
6
x

2 *Kinematics: Speed and Velocity*

Answers to Selected Discussion Questions

•2.1

If the average speed is zero, the object didn't move, so it is not possible for it to have a nonzero speed over a smaller interval. The average velocity may well be zero even for a moving body if it returned to its original position. Hence, it could have a nonzero value over a shorter interval.

•2.4

Yes. Imagine two people starting out at 9:00 a.m. on the same day, one at the top going down, the other at the bottom going up. They must meet at the same time somewhere.

•2.7

The dog is already running when the clock starts at $t = 0$ and it continues at a constant speed until $t = 2\,\text{s}$, at which point it stops, having gone $3\,\text{m}$. It rests until $t = 4\,\text{s}$ and then runs at a constant speed until $t = 10\,\text{s}$, whereupon it puts on a burst of speed. Nothing can be said about the displacement — the dog could be running in circles for all we know. It traveled a total distance of $6\,\text{m}$. It moved the fastest from $10\,\text{s}$ to $11\,\text{s}$.

•2.9

Changing the units will change the numerical value of the magnitude of a vector, but not its direction in space.

•2.10

The mouse began moving at $t = 1\,\text{s}$ and $s_x = 1\,\text{m}$. It ran to a point $2\,\text{m}$ from the origin in $1\,\text{s}$ and stopped. It rested for $1\,\text{s}$ and then, at $t = 2\,\text{s}$, it started running at nonuniform speed until it got 6 m from the origin. There, it immediately turned around and ran at a constant speed back to the opening of the tunnel.

•2.12

(a) At $t = 0$, $s = 0/DC = 0$. (b) $s \approx (At^2 + Bt)/Dt$. (c) $s \approx (At^2 + Bt)/DC$.

Answers to Odd-Numbered Multiple Choice Questions

1. d **3.** b **5.** d **7.** c **9.** c **11.** d **13.** a
15. c **17.** d **19.** d

Solutions to Selected Problems

2.11

(a) The length of each of the two segments of the trip is given by $l = 30\,\text{km}/2 = 15\,\text{km}$. Thus for the first segment

$$v_{av} = \frac{l}{t_1} = \frac{15\,\text{km}}{(15\,\text{min})(1.0\,\text{h}/60\,\text{min})} = 60\,\text{km/h}\,,$$

and for the second one

$$v_{av} = \frac{l}{t_2} = \frac{15\,\text{km}}{(45\,\text{min})(1.0\,\text{h}/60\,\text{min})} = 20\,\text{km/h}\,.$$

(b) For the whole trip the total distance covered is $l_{total} = 30\,\text{km}$, while the total time elapsed is $t = t_1 + t_2 = 15\,\text{min} + 45\,\text{min} = 60\,\text{min}$. Thus the average speed for the whole trip is

$$v_{av} = \frac{l_{total}}{t} = \frac{30\,\text{km}}{(60\,\text{min})(1.0\,\text{h}/60\,\text{min})} = 30\,\text{km/h}\,.$$

The distance vs time diagram is shown below. The average speeds for the first, second segment and the whole trip are represented by the slopes of lines a, b and c, respectively, in the plot in the next page.

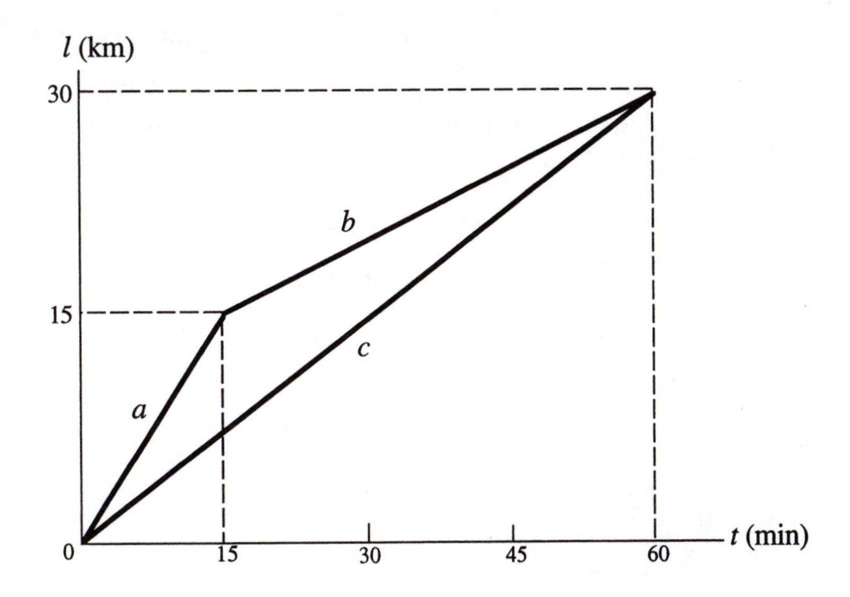

2.25

The speed v of the bee is a constant ($= 10\,\text{m/s}$), and its travel time is $t = 2.83\,\text{s}-1.33\,\text{s} = 1.50\,\text{s}$. Thus the distance traveled is given by $l = vt = (10\,\text{m/s})(1.50\,\text{s}) = 15\,\text{m}$.

2.35

Suppose that the radius of the remaining circular leaf is r at a given moment t. Then the area of the leaf at that instant is $A = \pi r^2$. As the caterpillars advance inward at a constant radial speed v, the radius of the leaf decreases at the rate of v: $dr/dt = -v$. Here the negative sign corresponds to the fact that r is *decreasing* with time. The resulting rate of change of the area of the leaf is

$$\frac{dA}{dt} = \frac{d}{dt}(\pi r^2) = \pi\frac{dr^2}{dt} = 2\pi r\frac{dr}{dt} = 2\pi r(-v) = -2\pi r v.$$

The rate at which the leaf is vanishing is therefore $dA/dt = -2\pi R v$, when $r = R$.

2.47

(a) The volume of a spherical balloon of radius r is $V = 4\pi r^3/3$. Since the radial speed of every point on the surface is v ($= 0.30\,\text{m/s}$) the radius of the balloon must be increasing at that rate: $dr/dt = v = 0.30\,\text{m/s}$. The corresponding rate at which V increases is then

$$\frac{dV}{dt} = \frac{d}{dt}\left(\frac{4\pi r^3}{3}\right) = \frac{4\pi}{3}\frac{dr^3}{dt} = \frac{4\pi}{3}\left(3r^2\frac{dr}{dt}\right) = 4\pi r^2\frac{dr}{dt} = 4\pi r^2 v.$$

(This result is expected. In fact, since $4\pi r^2$ is the surface area of the sphere $dV = 4\pi r^2 dr$, which yields the result above if we divide both sides by dt.) Now, since the radius grows from its initial value of zero at a constant speed v, $r = vt$. Thus dV/dt as a function of time is

$$\frac{dV}{dt} = 4\pi r^2 v = 4\pi(vt)^2 v = 4\pi v^3 t^2.$$

(b) Substitute $v = 0.30\,\text{m/s}$ and $t = 0.50\,\text{s}$ into the expression for dV/dt obtained above:

$$\left.\frac{dV}{dt}\right|_{t=0.50\,\text{s}} = 4\pi v^3 t^2\Big|_{t=0.50\,\text{s}} = 4\pi(0.30\,\text{m/s})^3(0.50\,\text{s})^2 = 8.5\times10^{-2}\,\text{m}^3/\text{s}\,.$$

2.51

Set up an xy coordinate system centered at the base of the wall, as shown, with the x axis lying horizontally and the y axis vertically upward, coinciding with the wall. The coordinates of the two ends of the ladder are $(0,\ y)$ (point A, on the wall) and $(x,\ 0)$ (point B, on the floor). Note that ΔAOB is right-angled; and so $\overline{OB}^2 + \overline{OA}^2 = \overline{AB}^2 = L^2$, where $L = 10.0\,\text{m}$ is the length of the ladder. Since $\overline{OB} = x$ and $\overline{OA} = y$, this is

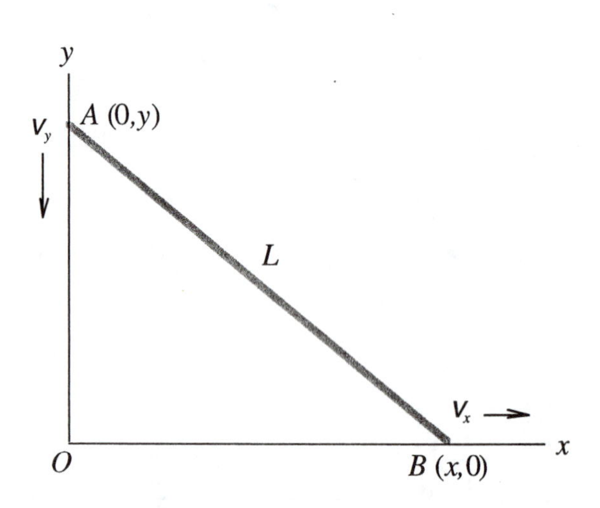

$$x^2 + y^2 = L^2\,.$$

As the base of the ladder (point B) is being pulled away at a constant speed v_x, the x-coordinate of point B increases at the rate of v_x: $dx/dt = v_x$. Meanwhile as point A (the top of the ladder) slides down the wall its y-coordinate changes at the rate of dy/dt, which is related to the speed v_y of point A by $v_y = -dy/dt$, where the negative sign reflects the fact that the top of the ladder is sliding *downward* (so $dy/dt < 0$) as the base of the ladder is pulled away from the wall. Taking the time derivative of the equation above, we obtain

$$\frac{d}{dt}(x^2+y^2) = 2x\frac{dx}{dt} + 2y\frac{dy}{dt} = 2xv_x - 2yv_y = \frac{dL^2}{dt} = 0\,,$$

which we solve for v_y:

$$v_y = \frac{xv_x}{y} = \frac{xv_x}{\sqrt{L^2-x^2}}\,,$$

where in the last step we used $x^2+y^2 = L^2$ to eliminate y.

(b) Substitute $x = 1.5\,\text{m}$, $v_x = 1.0\,\text{m/s}$, and $L = 10.0\,\text{m}$ into the expression for v_y above to obtain

$$v_y = -\frac{dy}{dt} = \frac{xv_x}{\sqrt{L^2-x^2}} = \frac{(1.5\,\text{m})(1.0\,\text{m/s})}{\sqrt{(10.0\,\text{m})^2-(1.5\,\text{m})^2}} = 0.15\,\text{m/s}\,.$$

(c) Refer to the expression for v_y obtained in part (a) above. As time progresses x increases (as the ladder is being pulled further to the right) while y decreases (as the top of the ladder slides closer to the base of the wall). Meanwhile, v_x remains a constant. Thus $v_y = v_x(x/y) \propto x/y$ increases, meaning that the top of the ladder slides with increasing speed as it descends towards the base of the wall.

2.53

The frog's displacement, $\vec{s}$, consists of two segments, $\vec{s}_1$ (vertically downward) and $\vec{s}_2$ (horizontally away from the table), as shown in the diagram to the left. Since $\vec{s}_1 \perp \vec{s}_2$, from the Pythagorean Theorem $s^2 = s_1^2 + s_2^2$; so

$$s = \sqrt{s_1^2 + s_2^2} = \sqrt{(1.0\,\text{m})^2 + (1.0\,\text{m})^2} = 1.4\,\text{m}\,.$$

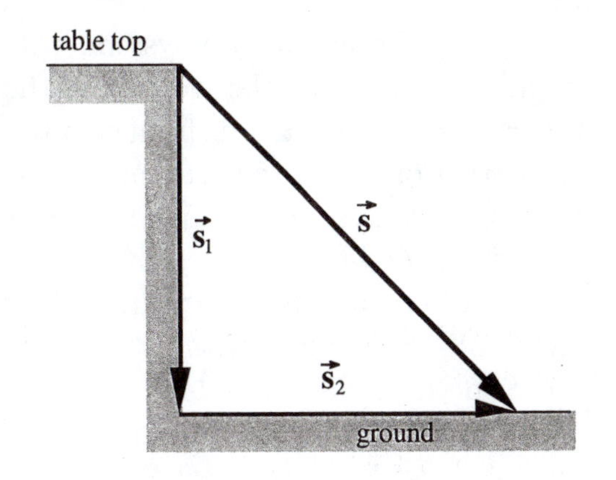

2.69

The magnitude of the horizontal displacement of the cannon ball is $s_x = 20\,\text{m}$, and that of the vertical displacement is $s_y = 60\,\text{m}$. Since $\vec{s}_x \perp \vec{s}_y$, the magnitude of the net displacement $\vec{s}$ is

$$s = \sqrt{s_x^2 + s_y^2} = \sqrt{(20\,\text{m})^2 + (60\,\text{m})^2} = 63\,\text{m}\,.$$

$\vec{s}$ is at an angle θ up from the horizontal, where θ satisfies

$$\tan\theta = \frac{s_y}{s_x} = \frac{60\,\text{m}}{20\,\text{m}} = 3.0\,.$$

The angle is $\theta = \tan^{-1}(3.0) = 72°$.

2.95

On the way up the bullet covers a distance of $l = 4588\,\text{m}$ in $t = 30.6\,\text{s}$, so

$$v_{\text{av}} = \frac{l}{t} = \frac{4588\,\text{m}}{30.6\,\text{s}} = 150\,\text{m/s}\,.$$

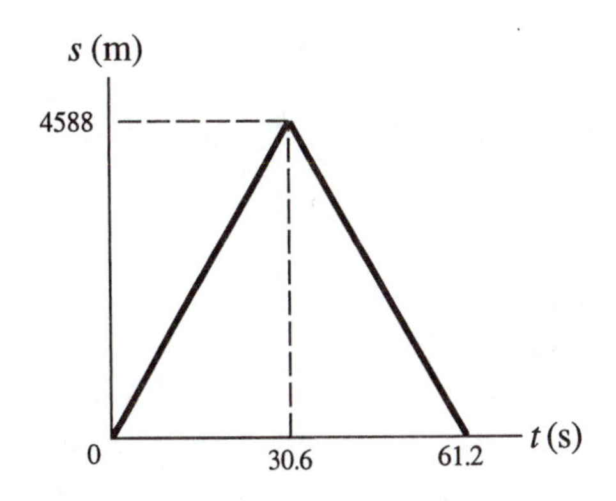

Since the displacement $\vec{s}$ and the distance l have the same magnitude on the way up, the velocity $\vec{v} = \vec{s}/t$ of the bullet also has a magnitude of 150 m/s, and is directed upward.
For the round trip the bullet travels twice as much distance as l in twice as much time as t, so the average speed remains 150 m/s. However, since $\vec{s} = 0$ for the round trip, $\vec{v}_{\text{av}} = \vec{s}/t_{\text{total}} = 0$.

2.115

Set up an xyz coordinate system, with its origin (O) coinciding the location of the car at $t = 0$, the x axis running from west to east, the y axis from south to north, and the z axis pointing vertically upward. The car starts off at point O when $t = 0$, moves along the positive x axis and reaches point C by time t, with $\overline{OC} = x = v_C t$, where $v_C = 10.0\,\text{m/s}$ is the speed of the car. Meanwhile the van starts from point A, located on the z axis a distance l_0 ($= 10.0\,\text{m}$) below point O, moves in the positive y direction and reaches point V by time t, with $\overline{AV} = y = v_V t$, where $v_V = 20.0\,\text{m/s}$ is the speed of the van. The separation between the car and the van is $l(t) = \overline{VC}$.

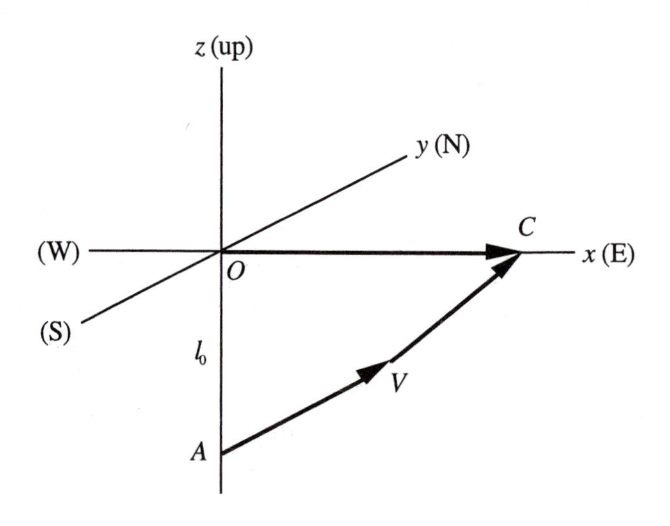

To find $l(t)$, note that the coordinates of C are $(x,\,0,\,0)$ and those for point V are $(0,\,y,\,-l_0)$; and so

$$l^2 = (x-0)^2 + (0-y)^2 + [0-(-l_0)]^2 = x^2 + y^2 + l_0^2 = (v_C t)^2 + (v_V t)^2 + l_0^2\,.$$

The time rate-of-change of $l(t)$ can be obtained by taking the time derivative of both sides of this equation:

$$\frac{dl^2}{dt} = 2l\frac{dl}{dt} = \frac{d}{dt}\left[(v_C t)^2 + (v_V t)^2 + l_0^2\right] = 2v_C^2 t + 2v_V^2 t\,,$$

or

$$\frac{dl}{dt} = \frac{2v_C^2 t + 2v_V^2 t}{2l} = \frac{(v_C^2 + v_V^2)t}{\sqrt{(v_C t)^2 + (v_V t)^2 + l_0^2}}\,,$$

where in the last step we used $l = \sqrt{(v_C t)^2 + (v_V t)^2 + l_0^2}$. Now plug in $t = 10.0\,\text{s}$, along with $v_V = 20.0\,\text{m/s}$, $v_C = 10.0\,\text{m/s}$, and $l_0 = 10.0\,\text{m}$ to obtain $dl/dt = 22.3\,\text{m/s}$.

2.119

Since both runners run at the same speed ($v = 5.00\,\text{m/s}$), they will meet at the midpoint of the track, covering $l = \frac{1}{2}(1000\,\text{m}) = 500\,\text{m}$ apiece. Thus the time t elapsed before they meet is $t = l/v = 500\,\text{m}/(5.00\,\text{m/s}) = 100\,\text{s}$. So the fly is in the air for 100 s, during which time it flies nonstop at a constant speed of $v = 10\,\text{m/s}$. Thus the path length it covers is

$$l = vt = (10\,\text{m/s})(100\,\text{s}) = 1.0 \times 10^3\,\text{m}\,.$$

The fact that the fly turns around many times is irrelevant to our result, as long as it spends a negligible amount of time doing so.

2.133

In the diagram shown to the right $\vec{\mathbf{s}}_{\mathrm{H}} = \vec{\mathbf{v}}_{\mathrm{H}}t$ is the displacement of the hawk (H), and $\vec{\mathbf{s}}_{\mathrm{M}} = \vec{\mathbf{v}}_{\mathrm{M}}t$ is that of the mouse (M). Here $|\vec{\mathbf{v}}_{\mathrm{H}}|$ is the desired speed of the hawk, $|\vec{\mathbf{v}}_{\mathrm{M}}| = 2.0\,\mathrm{m/s}$, and $t = 5.0\,\mathrm{s}$. Use the Pythagorean Theorem to obtain $H^2 + s_{\mathrm{M}}^2 = s_{\mathrm{H}}^2$, or $H^2 + (|\vec{\mathbf{v}}_{\mathrm{M}}|t)^2 = (|\vec{\mathbf{v}}_{\mathrm{H}}|t)^2$, from which we solve for $|\vec{\mathbf{v}}_{\mathrm{H}}|$:

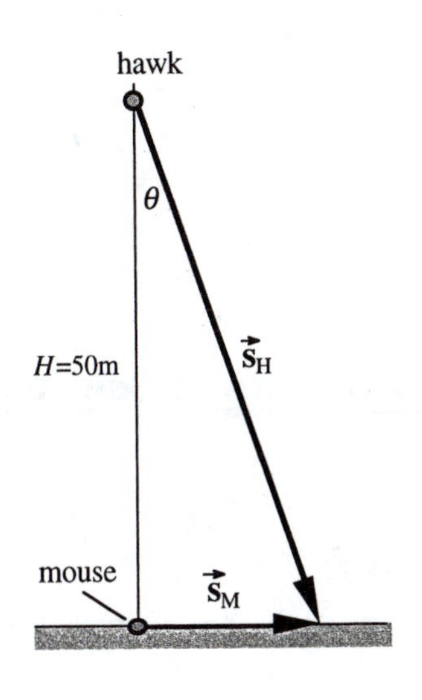

$$
\begin{aligned}
|\vec{\mathbf{v}}_{\mathrm{H}}| &= \frac{\sqrt{H^2 + |\vec{\mathbf{v}}_{\mathrm{M}}|^2 t^2}}{t} \\
&= \frac{\sqrt{(50\,\mathrm{m})^2 + [(2.0\,\mathrm{m/s})(5.0\,\mathrm{s})]^2}}{5.0\,\mathrm{s}} \\
&= 10\,\mathrm{m/s}\,.
\end{aligned}
$$

$\vec{\mathbf{v}}_{\mathrm{H}}$ should make an angle θ with respect to the vertical direction, where

$$
\tan\theta = \frac{s_{\mathrm{M}}}{H} = \frac{|\vec{\mathbf{v}}_{\mathrm{M}}|t}{H} = \frac{(2.0\,\mathrm{m/s})(5.0\,\mathrm{s})}{50\,\mathrm{m}} = 0.20\,.
$$

The angle is $\theta = \tan^{-1}(0.20) = 11°$.

2.139

The velocity of the ship (S) relative to the Earth (E), $\vec{\mathbf{v}}_{\mathrm{SE}}$, makes an angle of $30° + 60° = 90°$ with respect to $\vec{\mathbf{v}}_{\mathrm{JS}}$, the velocity of the jogger (J) relative to the ship. Since $\vec{\mathbf{v}}_{\mathrm{JE}} = \vec{\mathbf{v}}_{\mathrm{JS}} + \vec{\mathbf{v}}_{\mathrm{SE}}$ and $\vec{\mathbf{v}}_{\mathrm{JS}} \perp \vec{\mathbf{v}}_{\mathrm{SE}}$, from the Pythagorean Theorem we have $|\vec{\mathbf{v}}_{\mathrm{JE}}|^2 = |\vec{\mathbf{v}}_{\mathrm{JS}}|^2 + |\vec{\mathbf{v}}_{\mathrm{SE}}|^2$. Solve for $|\vec{\mathbf{v}}_{\mathrm{JE}}|$:

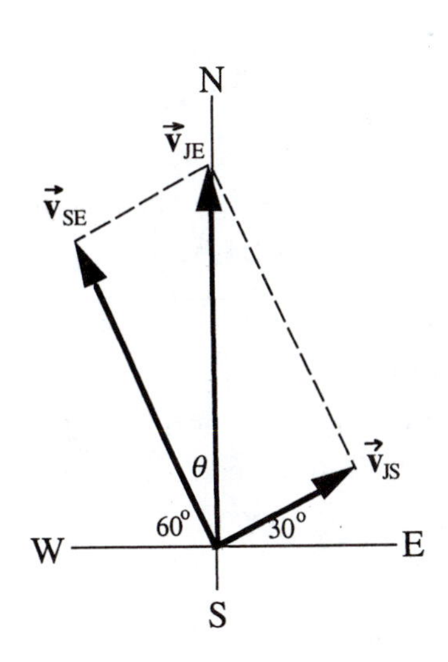

$$
\begin{aligned}
|\vec{\mathbf{v}}_{\mathrm{JE}}| &= \sqrt{|\vec{\mathbf{v}}_{\mathrm{JS}}|^2 + |\vec{\mathbf{v}}_{\mathrm{SE}}|^2} \\
&= \sqrt{(10\,\mathrm{km/h})^2 + (20\,\mathrm{km/h})^2} = 22\,\mathrm{km/h}\,.
\end{aligned}
$$

The angle θ which $\vec{\mathbf{v}}_{\mathrm{JE}}$ makes with the direction of motion of the ship satisfies

$$
\tan\theta = \frac{|\vec{\mathbf{v}}_{\mathrm{JS}}|}{|\vec{\mathbf{v}}_{\mathrm{SE}}|} = \frac{10\,\mathrm{km/h}}{20\,\mathrm{km/h}} = 0.50\,,
$$

which gives $\theta = 27°$. Thus $\vec{\mathbf{v}}_{\mathrm{JE}}$ is at an angle of $30° - 27° = 3°$ west of north.

3 Kinematics: Acceleration

Answers to Selected Discussion Questions

•3.3

The balls are identical and separately must fall together, that is, at the same rate. When two are stuck together and dropped alongside a single ball, they again must fall together provided the air resistance on the double mass is no different. In vacuum, there is no air resistance so both chunks of clay (one twice the mass of the other) must fall together. Indeed, 1000 such balls of clay stuck together and shaped into a piano would fall the same way all things must fall together in vacuum.

•3.7

The cars are always separated by one second. The distance each travels is proportional to the time squared so their separation will increase: after 1 s, the separation will be $\frac{1}{2}a(1\,\mathrm{s})^2$; after 10 s, it will be $\frac{1}{2}a[(10\,\mathrm{s})^2 - (9\,\mathrm{s})^2] = \frac{1}{2}a(19\,\mathrm{s}^2)$.

•3.9

If the acceleration were constant $v_{\mathrm{av}} = \frac{1}{2}(v_0 + v)$, and the average is midway between the initial and final speeds. But with a increasing, the average speed is closer to the initial speed than to the final speed.

•3.11

The keys fall at the same rate as the floor and so cannot get any closer. The keys float in front of your face, which is also falling at the same rate.

•3.15

(a) The acceleration a as a function of time t is given by

$$a = \frac{(At^2 - Bt)}{(t + C)D} + \frac{Et^2}{(t - C)D}\,.$$

When $t = 0$, $a = 0/CD + 0/(-CD) = 0$.
(b) When $t \gg C$, $t \pm C \approx t$; so

$$a \approx \frac{At^2 - Bt}{tD} + \frac{Et^2}{tD} = \frac{At - B + Et}{D}.$$

(c) When $t \ll C$, $t \pm C \approx \pm C$; so

$$a \approx \frac{At^2 - Bt}{CD} - \frac{Et^2}{CD} = \frac{At^2 - Bt - Et^2}{CD}.$$

•3.17
The speed reaches a maximum of 4.2 s and then drops to zero at the peak altitude ($t_p \approx 7.4$ s). The net acceleration increases for about 2 s, after which it decreases as the engine throttles down. At $t = 4.2$ s, with the engine still firing, the net acceleration is zero. Thereafter it is increasingly negative as the engine's thrust decreases and gravity dominates. The engine shuts off at around 5.5 s and the rocket decelerates moving upward until it stops at $t \approx 7.4$ s.

Answers to Odd-Numbered Multiple Choice Questions

1. b **3.** d **5.** a **7.** a **9.** b **11.** d **13.** a
15. c **17.** d

Solutions to Selected Problems

3.1
The initial speed of the rocket is $v_i = 0$ and the final speed is $v_f = 100$ m/s. Thus from Eq. (3.1) the acceleration is given by

$$a_{av} = \frac{v_f - v_i}{\Delta t} = \frac{100\,\text{m/s} - 0}{10\,\text{s}} = 10\,\text{m/s}^2.$$

3.5

Now the initial speed is $v_i = 1.0\,\text{m/s}$, the final speed is $v_f = 15.0\,\text{m/s}$, and the time interval is given by $\Delta t = 1\,\text{min}\,2\,\text{s} = 60\,\text{s} + 2\,\text{s} = 62\,\text{s}$. Thus from Eq. (3.1) the acceleration is found to be

$$a_{av} = \frac{v_f - v_i}{\Delta t} = \frac{15.0\,\text{m/s} - 1.0\,\text{m/s}}{62\,\text{s}} = 0.23\,\text{m/s}^2\,.$$

3.17

The acceleration a can be obtained from the slope of the velocity vs time graph. Thus from Fig. (3.2a)

$$a_{av} = \text{slope} = \frac{\text{rise}}{\text{run}} = \frac{25.0\,\text{m/s}}{5.0\,\text{s}} = 5.0\,\text{m/s}^2\,.$$

3.43

The length of the path traversed by the vehicle (without explicit units) is $l(t) = 12t^3 - 6.0t^2 + 2.4t$. The corresponding speed is

$$v(t) = \frac{dl}{dt} = \frac{d}{dt}\left(12.0\,t^3 - 6.0t^2 + 2.4t\right) = 36.0\,t^2 - 12t + 2.4\,,$$

while its tangential acceleration as a function of time is

$$a_t(t) = \frac{dv}{dt} = \frac{d}{dt}(36.0\,t^2 - 12t + 2.4) = 72.0\,t - 12 = (72.0\,\text{m/s}^3)t - 12\,\text{m/s}^2\,.$$

Set $a_t(t) = 0$ to obtain $t = (12\,\text{m/s}^2)/(72.0\,\text{m/s}^3) = \frac{1}{6}\,\text{s} = 0.17\,\text{s}$.

3.75

First, use $v^2 - v_0^2 = 2as$ to find the acceleration a of the swimmer as she slows down from $v_0 = 2.2\,\text{m/s}$ to $v = 0$ over a distance of $s = 10\,\text{m}$:

$$a = \frac{v^2 - v_0^2}{2s} = \frac{0 - (2.2\,\text{m/s})^2}{2(10\,\text{m})} = -0.242\,\text{m/s}^2\,.$$

Now, at this acceleration, she can cover a distance of $s = v_0 t + \frac{1}{2}at^2$ a time t after the onset of the acceleration. Therefore at the beginning of the 3rd second (i.e., the end of the 2nd second) her displacement is given by $s_2 = v_0 t_2 + \frac{1}{2}at_2^2$, where $t_2 = 2.0\,\text{s}$; while at the end of the 3rd second it becomes $s_3 = v_0 t_3 + \frac{1}{2}at_3^2$, where $t_3 = 3.0\,\text{s}$. Her net displacement during the third second is thus

$$\begin{aligned}\Delta s = s_3 - s_2 &= \left(v_0 t_3 + \frac{1}{2}at_3^2\right) - \left(v_0 t_2 + \frac{1}{2}at_2^2\right) = v_0(t_3 - t_2) + \frac{1}{2}a(t_3^2 - t_2^2)\\ &= (2.2\,\text{m/s})(3.0\,\text{s} - 2.0\,\text{s}) + \frac{1}{2}(-0.242\,\text{m/s}^2)\left[(3.0\,\text{s})^2 - (2.0\,\text{s})^2\right]\\ &= 1.6\,\text{m}\,.\end{aligned}$$

3.79

(a) By the time the second missile is just launched the first one has already been in acceleration (at a rate of $a = 20\,\mathrm{m/s^2}$) for $t = 1.0\,\mathrm{s}$. Thus the separation between the two is

$$\Delta s = v_0 t + \frac{1}{2}at^2 = (60\,\mathrm{m/s})(1.0\,\mathrm{s}) + \frac{1}{2}(20\,\mathrm{m/s^2})(1.0\,\mathrm{s})^2 = 70\,\mathrm{m}\,.$$

(b) At the moment the third missile is launched, the second one has just spent 1.0 s in the air, so it has moved a distance of $s_2 = 70\,\mathrm{m}$ (see the calculation above). Meanwhile, the first missile has been in the air for $t = 2.0\,\mathrm{s}$, so its displacement from the launch point is now

$$s_1 = v_0 t + \frac{1}{2}at^2 = (60\,\mathrm{m/s})(2.0\,\mathrm{s}) + \frac{1}{2}(20\,\mathrm{m/s^2})(2.0\,\mathrm{s})^2 = 160\,\mathrm{m}\,.$$

Thus the separation between the first and the second missile is now $\Delta s = s_1 - s_2 = 160\,\mathrm{m} - 70\,\mathrm{m} = 90\,\mathrm{m}$.

3.95

Your initial speed v_0, displacement s, final speed v and acceleration g are related by $v^2 - v_0^2 = 2gs$. Since $v_0 = 0$,

$$v = \sqrt{2gs} = \sqrt{2(9.81\,\mathrm{m/s^2})(0.50\,\mathrm{m})} = 3.1\,\mathrm{m/s}\,.$$

Use $1\,\mathrm{ft} = 0.3048\,\mathrm{m}$ to convert v to ft/s: $v = (3.1\,\mathrm{m/s})(1\,\mathrm{ft}/0.3048\,\mathrm{m}) = 10\,\mathrm{ft/s}$. Use $1\,\mathrm{mi} = 1609\,\mathrm{m}$ and $1\,\mathrm{h} = 3600\,\mathrm{s}$ to convert v to mi/h: $v = (3.13\,\mathrm{m/s})(1\,\mathrm{mi}/1609\,\mathrm{m})(3600\,\mathrm{s/h}) = 7.0\,\mathrm{mi/h}$.

3.127

If the velocity $\vec{v}$ of the projectile at a time t has a horizontal component of v_x and a vertical component of v_y, then the angle θ_t which $\vec{v}$ makes with the horizontal direction is given by $\tan\theta_t = v_y/v_x$, or $\theta_t = \tan^{-1}(v_y/v_x)$, as shown in the figure to the right. Now, the horizontal part of any projectile motion is uniform so at any time t into its flight $v_x = v_{0x} = v_0\cos\theta$. Vertically, the motion has a uniform acceleration g so at time t we have $v_y = v_{0y} + gt = v_0\sin\theta + gt$. Thus

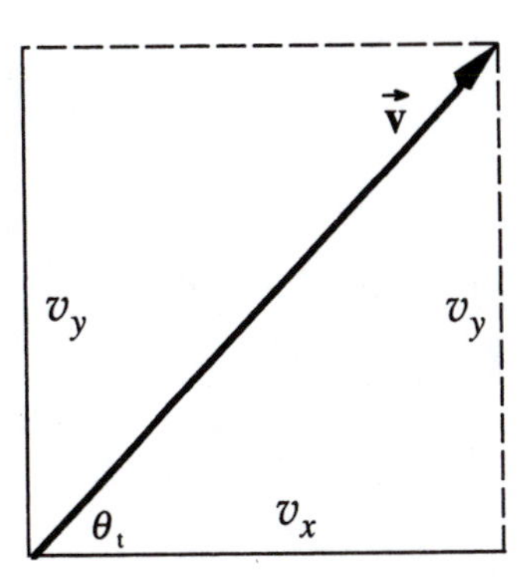

$$\theta_t = \tan^{-1}\left(\frac{v_y}{v_x}\right) = \tan^{-1}\left(\frac{v_0\sin\theta + gt}{v_0\cos\theta}\right) = \tan^{-1}\left(\tan\theta + \frac{gt}{v_0\cos\theta}\right),$$

where the trigonometric identity $\sin\theta/\cos\theta = \tan\theta$ was used.

3.133

For simplicity of notation, let's temporarily suppress the unit and ignore the significant figure notation in the expression for the speed of the object: $v(t) = t^2 - t - 6$. Now integrate $v(t)$ over the time interval between $t_0 = 1.0\,\mathrm{s}$ and $t = 4.0\,\mathrm{s}$ to find the displacement of the object during that interval [see Eq. (3.23)]:

$$s(t) - s(t_0) = \int_{t_0}^{t} v(t)\,dt = \int_{1.0}^{4.0} (t^2 - t - 6)\,dt = \left[\frac{1}{3}t^3 - \frac{1}{2}t^2 - 6t\right]_{1.0}^{4.0} = -4.5\,\mathrm{m}\,.$$

Note that in the last step we restored the unit of the displacement and retained two significant figures, in accordance with the significant figures given in $v(t)$, t_0 and t.

3.141

Since we are looking for the total *distance*, rather than the displacement, we need to be careful about the possibility that $v(t)$ may change sign during the time interval $1.00\,\mathrm{s} < t < 4.00\,\mathrm{s}$, in which case the distance traveled exceeds the magnitude of the net displacement. Indeed, if we drop the units for the moment and write

$$v(t) = t^2 - t - 6 = (t+2)(t-3)\,,$$

then it is clear that $v(t)$ is negative for $1.00\,\mathrm{s} \le t < 3.00\,\mathrm{s}$, and positive for $3.00\,\mathrm{s} < t \le 4.00\,\mathrm{s}$. So we need to compute separately the distance traveled in these two time segments. For $1.00\,\mathrm{s} \le t < 3.00\,\mathrm{s}$, the displacement of the object is

$$s_1 = \int_{1.00}^{3.00} v(t)\,dt = \int_{1.00}^{3.00} (t^2 - t - 6)\,dt = \left[\frac{1}{3}t^3 - \frac{1}{2}t^2 - 6t\right]_{1.00}^{3.00} = -7.333\,\mathrm{m}\,,$$

and so the distance it travels during that time period is $l_1 = |s_1| = 7.33\,\mathrm{m}$. For $3.00\,\mathrm{s} \le t < 4.00\,\mathrm{s}$, the displacement of the object is

$$s_2 = \int_{3.00}^{4.00} (t^2 - t - 6)\,dt = \left[\frac{1}{3}t^3 - \frac{1}{2}t^2 - 6t\right]_{3.00}^{4.00} = 2.83\,\mathrm{m}\,,$$

and so the distance it travels during that time period is $l_2 = |s_2| = 2.83\,\mathrm{m}$. The total distance traveled is then

$$l_{\mathrm{total}} = l_1 + l_2 = 7.33\,\mathrm{m} + 2.83\,\mathrm{m} = 10.2\,\mathrm{m}\,.$$

4 *Newton's Three Laws*

Answer to Selected Discussion Question

•4.2
His ship was already moving rapidly so he should have shut off the engines and coasted, conserving fuel for the landing when he must change velocity. The writers didn't know the Law of Inertia.

Answers to Odd-Numbered Multiple Choice Questions

1. d	**3.** e	**5.** c	**7.** c	**9.** d	**11.** c	**13.** d
15. a	**17.** a	**19.** a	**21.** d	**22.** b	**23.** c	

Solutions to Selected Problems

4.3
The vertical separation between the grapes and the open mouth of Marc Antony is $s_y = 1.0000\,\mathrm{m}$. Since the grapes are released with no initial vertical velocity, their time-of-flight t satisfies $s_y = \frac{1}{2}gt^2$, or $t = \sqrt{2s_y/g}$. In the mean time, the grapes must move horizontally by

s_x at a constant speed of $v_x = 2.2136\,\text{m/s}$, so they must be released a horizontal distance s_x from him, where

$$s_x = v_x t = v_x \sqrt{\frac{2s_y}{g}} = (2.2136\,\text{m/s})\sqrt{\frac{2(1.0000\,\text{m})}{9.8000\,\text{m/s}^2}} = 1.0000\,\text{m}\,.$$

4.13

The monkey and the dart start their respective motion at the same instant. Note that the monkey falls vertically so its horizontal position does not change. Suppose that it takes a time t for the dart to close the initial horizontal separation between itself and the monkey. Now, if there were no gravity, then the dart would just move along the line-of-sight. Due to the presence of gravity, however, the actual trajectory of the dart falls *below* the original line-of-sight by an amount $\frac{1}{2}gt^2$ by the time the dart reaches the same horizontal position as the monkey. The monkey, meanwhile, also falls vertically from the line-of-sight by exactly the same amount. So after a time t into the flight the dart and the monkey will be at the same *vertical* as well as *horizontal* position. This is why the dart will get the monkey.

4.17

Consider the three-force arrangement shown to the right. Each force has a magnitude of 2 kN. Since the net force is due east with no south- or north-component, we set the angle between $\vec{\mathbf{F}}_1$ and the x-axis to be the same as that between $\vec{\mathbf{F}}_2$ and the x-axis (denoted as θ in the figure to the right). Taking east as positive, we require that the sum of the east-west (i.e. x-) components of the three forces be +4 kN:

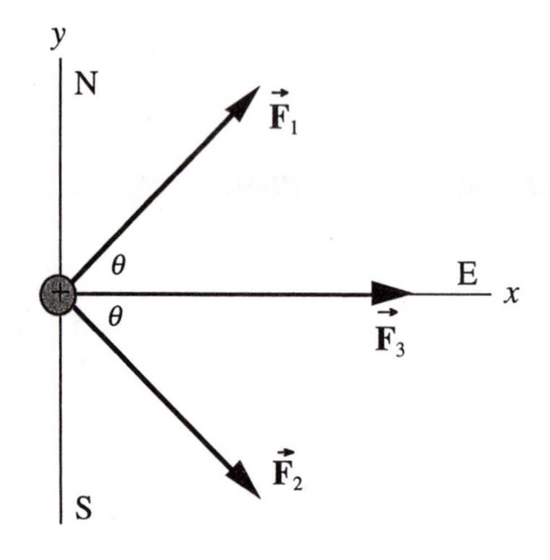

$$\overset{+}{\rightarrow}\sum F_x = 2(2\,\text{kN})\cos\theta + 2\,\text{kN} = 4\,\text{kN}\,,$$

which gives $\cos\theta = 1/2$. The angle is $\theta = 60°$.

4.26

Assuming the acceleration $\vec{\mathbf{a}}$ of the lion of mass m to be horizontal, from $\sum\vec{\mathbf{F}} = m\vec{\mathbf{a}}$ we know that $\vec{\mathbf{a}}$ must result from a horizontal force exerted on the lion by the ground. According to Newton's Third Law the lion must exert a reactive force of the same magnitude on the ground. The magnitude F of the force from the lion is therefore

$$F = ma = (170\,\text{kg})(10\,\text{m/s}^2) = 1.7 \times 10^3\,\text{N} = 1.7\,\text{kN}\,.$$

4. 31

The force $F(t)$ exerted on an object of mass m is related to the resulting acceleration $a(t)$ of the object via Newton's Second Law: $F(t) = m\,a(t)$. In our case we are given $v(t)$, from which we may obtain $a(t)$ as $a(t) = dv/dt$. Thus

$$F(t) = m\,a(t) = m\frac{dv}{dt} = m\frac{d}{dt}(3t^2 - 12t) = m(6t - 12)\,,$$

which is also in its proper SI unit (N).

4.47

The speed $v(t)$ of the asteroid can be found from its initial value v_0 ($= 2.00\,\mathrm{m/s}$), and its acceleration $a(t)$, as $v(t) = v_0 + \int_0^t a(t)\,dt$. Here $a(t)$ follows from the force $F(t)$ exerted on the asteroid (of mass m): $a(t) = F(t)/m = (At + B)/m$, with $A = 120 \times 10^2\,\mathrm{N/s}$ and $B = 40.0 \times 10^2\,\mathrm{N}$. Combining these relations, we get

$$\begin{aligned} v(t) &= v_0 + \int_0^t a(t)\,dt = v_0 + \frac{1}{m}\int_0^t (At + B)\,dt = v_0 + \frac{1}{m}\left(\frac{1}{2}At^2 + Bt\right) \\ &= 2.00\,\mathrm{m/s} + \frac{1}{100.0\,\mathrm{kg}}\left[\frac{1}{2}(120 \times 10^2\,\mathrm{N/s})t^2 + (40.0 \times 10^2\,\mathrm{N})t\right] \\ &= (60.0\,\mathrm{m/s^3})t^2 + (40.0\,\mathrm{m/s^2})t + 2.00\,\mathrm{m/s}\,. \end{aligned}$$

4.54

The downward weight of the frog of mass m is $F_{\mathrm{w}} = mg = (0.50\,\mathrm{kg})(9.8\,\mathrm{m/s^2}) = 4.9\,\mathrm{N}$. To lift the frog up one must apply an upward force whose magnitude exceeds 4.9 N. When a minimum force (barely greater than 4.9 N) is applied the net force on the frog is virtually zero, meaning that the frog can only be picked up at a very slow and constant speed.

4.65

Parallel to the incline, the net force exerted on the car is $\sum F_{\parallel} = F_{\mathrm{w}}\sin\theta = mg\sin\theta$, where m is the mass of the car and $\theta = 20^\circ$ is the angle of inclination. Set $\sum F_{\parallel} = ma$ to obtain the acceleration a of the car down the incline: $a = g\sin\theta$. The speed v of the car after sliding down by a distance of $s = 20\,\mathrm{m}$ is then given by $v^2 = 2as$, or

$$v = \sqrt{2as} = \sqrt{2gs\sin\theta} = \sqrt{2(9.81\,\mathrm{m/s^2})(20\,\mathrm{m})(\sin 20^\circ)} = 12\,\mathrm{m/s}\,.$$

4.67

In the figure shown to the right the angle in question is denoted as θ. The pom-pom of mass m has no vertical acceleration and a horizontal acceleration of a. Thus the net vertical force exerted on it must vanish:

$$+\uparrow \sum F_y = F_T \cos\theta - F_W = 0\,,$$

where $F_W = mg$ and F_T is the tension in the string; while the net horizontal force satisfies

$$\overset{+}{\rightarrow} \sum F_x = F_T \sin\theta = ma\,.$$

Rewrite the equation for $\sum F_y$ as $F_T \cos\theta = F_W = mg$, which we now use to divide both sides of the equation for $\sum F_x$. This yields

$$\frac{F_T \sin\theta}{F_T \cos\theta} = \tan\theta = \frac{ma}{mg} = \frac{a}{g}\,,$$

or $\tan\theta = a/g$. For $a = \Delta v/\Delta t = (26.8\,\mathrm{m/s})/6.8\,\mathrm{s} = 3.94\,\mathrm{m/s^2}$,

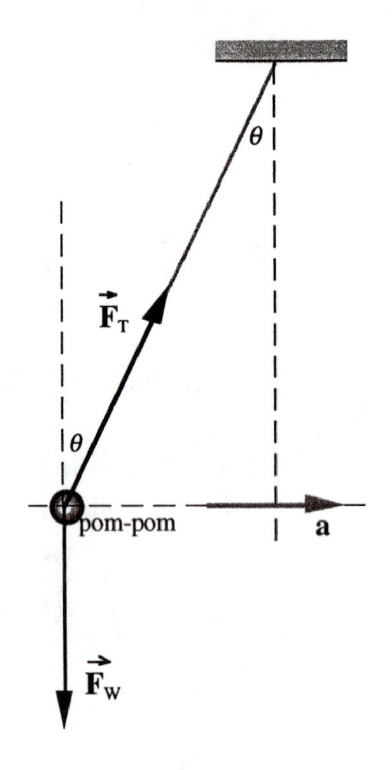

$$\theta = \tan^{-1}\left(\frac{a}{g}\right) = \tan^{-1}\left(\frac{3.94\,\mathrm{m/s^2}}{9.81\,\mathrm{m/s^2}}\right) = 22^\circ\,.$$

4.88

Refer to the free-body diagrams for the two masses shown in Fig. P86, with g replaced by g_M. Since $m_2 > m_1$ mass 1 will move up while mass 2 will move down. Taking up as positive and denoting the common magnitude of the accelerations of both masses as a and the tension in the rope as F_T, then for mass 1 $+\uparrow \sum F_{y1} = F_T - m_1 g_M = m_1 a$, and for mass 2 $+\uparrow \sum F_{y2} = F_T - m_2 g_M = m_2(-a) = -m_2 a$. Subtract the second equation from the first one to obtain $m_2 g_M - m_1 g_M = m_1 a + m_2 a$, which gives

$$g_M = \left(\frac{m_1 + m_2}{m_2 - m_1}\right) a\,.$$

Here $m_1 = 0.25\,\mathrm{kg}$, and $m_2 = 0.25\,\mathrm{kg} + 0.025\,\mathrm{kg} = 0.275\,\mathrm{kg}$ is the more massive of the two masses (the one with the gronch).

To find g_M we first need to find a. With the gronch in place, the system moves by $s_h = 0.50\,\mathrm{m}$ while its speed increases from 0 to v, at which time the gronch is removed. Thus $v^2 = 2as_h$.

Now, after the gronch is removed the acceleration is zero, and the speed v of the system is $v = 1.2\,\text{m}/3.0\,\text{s} = 0.40\,\text{m/s}$. This gives

$$a = \frac{v^2}{2s_{\text{h}}} = \frac{(0.40\,\text{m/s})^2}{2(0.50\,\text{m})} = 0.16\,\text{m/s}^2\,.$$

Substitute the value for a into the equation above for g_{M}:

$$g_{\text{M}} = \left(\frac{m_1 + m_2}{m_2 - m_1}\right)a = \left(\frac{0.25\,\text{kg} + 0.275\,\text{kg}}{0.275\,\text{kg} - 0.25\,\text{kg}}\right)(0.16\,\text{m/s}^2) = 3.4\,\text{m/s}^2\,.$$

4.93

Use Eq. (4.12), $\theta_{\text{max}} = \tan^{-1}\mu_{\text{s}}$, which gives $\mu_{\text{s}} = \tan\theta_{\text{max}}$. Here $\theta_{\text{max}} = 17°$ so $\mu_{\text{s}} = \tan 17° = 0.31$.

4.99

First, find the minimum stopping distance $s_x(\text{min})$ of the car. The initial speed of the car before deceleration is v_0. After moving forward by s_x its final speed is $v = 0$. This requires an acceleration a, which satisfies $v^2 - v_0^2 = -v_0^2 = 2a_x$, or $s_x = -v_0^2/2a_x$. Now, a is being provided by the force of static friction F_f between the car and the road. As F_f reaches its maximum value, so does a: $F_f(\text{max}) = -\mu_{\text{s}} F_{\text{N}} = \mu_{\text{s}} mg = ma_x(\text{max})$. (Here the minus sign indicates that the force is against the direction of motion so $a < 0$.) This gives $a_x(\text{max}) = -\mu_{\text{s}} g$. Substitute this result into the expression for s_x, the stopping distance:

$$s_x(\text{min}) = -\frac{v_0^2}{2a_x(\text{max})} = -\frac{v_0^2}{2(-\mu_{\text{s}} g)} = \frac{v_0^2}{2\mu_{\text{s}} g}\,.$$

Now, the average speed of the car during the deceleration process is $v_{\text{av}} = \frac{1}{2}(v_0 + v) = \frac{1}{2}v_0$ (as $v = 0$). Thus the minimum stopping time is

$$t_{\text{min}} = \frac{s_x(\text{min})}{v_{\text{av}}} = \frac{v_0^2}{2\mu_{\text{s}} g(v_0/2)} = \frac{v_0}{\mu_{\text{s}} g} = \frac{27\,\text{m/s}}{0.9(9.81\,\text{m/s}^2)} = 3\,\text{s}\,.$$

4.107

In example [4.9] in the text we learned $F_f = F_{\text{W}} \sin\theta$, where θ is the angle of inclination. To find θ_{max}, we need to first find $F_f(\text{max})$, for which we need to know F_{N}.

Normally, $F_{\text{N}} = F_{\text{W}} \cos\theta$, where F_{W} is the total weight of the object on the incline. In this case since the static friction from the road which pushes the vehicle is exerted only on the rear wheels, we should only count the part of F_{N} which falls on the rear wheels. Call this part F_{NR}. Then $F_{\text{NR}} = F_{\text{WR}} \cos\theta$, where $F_{\text{WR}} = (1 - 57\%)F_{\text{W}} = 43\% F_{\text{W}}$ is the portion of the weight of the vehicle that falls on the rear axle. The maximum static friction the vehicle can get from the road is then $F_f(\text{max}) = \mu_s F_{NR} = \mu_s(43\% F_W \cos\theta)$, which occurs when the angle of inclination is $\theta = \theta_{max}$, the steepest possible: $F_f(\text{max}) = 43\%\mu_s F_W \cos\theta = F_W \sin\theta_{max}$, which gives

$$\tan\theta_{max} = \frac{\sin\theta_{max}}{\cos\theta_{max}} = \frac{43\%\mu_s F_W}{F_W} = 43\%\mu_s = (0.43)(0.9) = 0.387\,.$$

The steepest angle is therefore $\theta_{max} = \tan^{-1}(0.387) = 21°$, or, to one significant figure, 0.2×10^2 degrees.

4.119

The tension F_{T1} in the rope connected to mass 1 has to support its weight. So $F_{T1} = F_{W1} = 15.0\,\text{N}$. Similarly, the tension F_{T2} in the rope connected to mass 2 is $F_{T2} = F_{W2} = 31.0\,\text{N}$. Now let's consider the balance of forces for the ring which is attached to both ropes as well as mass 3. Horizontally,

$$\overset{+}{\rightarrow}\sum F_x = F_{T2}\sin\phi - F_{T1}\sin\theta = 0\,;$$

and vertically

$$+\!\uparrow\sum F_y = F_{T2}\cos\phi + F_{T1}\cos\theta - F_{W3} = 0\,.$$

With $\theta = 45.0°$, $\phi = 20.0°$, $F_{T1} = 15.0\,\text{N}$ and $F_{T2} = 31.0\,\text{N}$, the equation for F_x is already satisfied (as you can easily check). Now solve for F_{W3} from the equation for F_y:

$$F_{W3} = F_{T2}\cos\phi + F_{T1}\cos\theta = (31.0\,\text{N})(\cos 20.0°) + (15.0\,\text{N})(\cos 45°) = 39.7\,\text{N}\,.$$

4.125

Let the tension in the rope be F_T. Note that the tension has to support the weight of each 1.0-kg mass so $F_T = (1.0\,\text{kg})(9.8\,\text{m/s}^2) = 9.8\,\text{N}$. Now consider the midpoint of the rope where the hook is located. By symmetry the net horizontal force exerted on it is zero. In the vertical direction, this point is subject to a downward force of $F_W = mg$ due to the weight of mass m, while it is being pulled up by the rope on both sides with a total upward force of $2F_T \sin 20°$. Thus for the midpoint

$$+\!\uparrow\sum F_y = 2F_T\sin 20° - F_W = 0\,,$$

which yields

$$F_W = 2F_T\sin 20° = 2(9.8\,\text{N})(\sin 20°) = 6.7\,\text{N}\,.$$

5 *Centripetal Force & Gravity*

Answers to Selected Discussion Questions

•5.1

An object can move in a direction other than that of the net force; for example, a ball flying through the air. No, it accelerates only in the direction of the net force.

•5.7

The Earth rotates counterclockwise looking down onto the North Pole, which means that any rocket on the launch pad is already moving eastward at the speed of the Earth even before lift-off. The closer to the equator, the greater the speed of a point on the Earth's surface. So, launch from a southerly location and launch eastward and you get as much benefit from the planet's rotation as possible.

•5.9

The rocket fires with a forward thrust and that carries it into an elongated elliptical orbit. The additional speed allows it to move outward beyond the circular orbit. It slows down as it moves away from the center of force. As it reaches its most distant point (just before it would begin to fall back inward), the rocket fires again, giving it enough speed to satisfy Eq. (5.11) and go into a large circular orbit. Despite all maneuvering, the final orbital speed is less than the initial orbital speed.

•5.13

These antennas pick up TV signals from satellites parked in geosynchronous orbits. In the Northern Hemisphere, these are all seen in the southern part of the sky since they are over the equator. Each craft is fixed at a known location in space so you need only decide which one you would like to tune in.

•5.15

Any point on the plane of the orbit is at distances from the Sun and Earth such that the magnitude of the gravitational attraction to each celestial body is the same. The resultant of the two opposing forces will therefore only have a component in that perpendicular plane, and it will always point toward the Earth-Sun axis. Thus, it will act as a central force, and the satellite will be held in orbit as if there were a central gravitating body.

Answers to Odd-Numbered Multiple Choice Questions

1. a **3.** b **5.** c **7.** d **9.** c **11.** c **13.** d
15. c **17.** b

Solutions to Selected Problems

5.19

At the top of the circular loop the person of mass m in circular motion is subject to two downward forces: his weight, mg; and F_N, which according to the problem statement is half of mg. Since F_c at this point is also vertically downward, we have

$$+\downarrow \sum F = mg + F_N = mg + \frac{1}{2}mg = F_c = m\frac{v^2}{r},$$

where v is the speed of the person and r is the radius of the circular loop. Solve for v:

$$v = \sqrt{\frac{3}{2}gr} = \sqrt{\frac{3}{2}(9.81\,\text{m/s}^2)(25.0\,\text{m})} = 19.2\,\text{m/s}.$$

5.31

According to Newton's Law of Universal Gravitation the gravitational force between two objects of mass m and M, separated by a distance of r, is $F_G = GmM/r^2$ [see Eq. (5.5)]. The gravitational attraction between Uranus (U) and Neptune (N) is then given by

$$\begin{aligned} F_{\mathrm{G}} &= \frac{GM_{\mathrm{U}}M_{\mathrm{N}}}{r_{\mathrm{UN}}^2} = \frac{G(14.6M_{\oplus})(17.3M_{\oplus})}{r_{\mathrm{UN}}^2} \\ &= \frac{(6.67\times10^{-11}\,\mathrm{N{\cdot}m^2/kg^2})(14.6)(17.3)(5.975\times10^{24}\,\mathrm{kg})^2}{(4.9\times10^{12}\,\mathrm{m})^2} \\ &= 2.5\times10^{16}\,\mathrm{N}\,. \end{aligned}$$

5.45

From Eq. (5.6) $g_0 = GM_{\oplus}/R_{\oplus}^2$, and from Eq. (5.7) $g_{\oplus} = GM_{\oplus}/r^2$. Thus

$$g_{\oplus}(r) = \frac{GM_{\oplus}}{r^2} = \left(\frac{GM_{\oplus}}{R_{\oplus}^2}\right)\left(\frac{R_{\oplus}}{r}\right)^2 = g_0\left(\frac{R_{\oplus}}{r}\right)^2.$$

5.59

The mass of a spherical neutron star of density ρ_{n} and radius R is given by $M_{\mathrm{n}} = \rho_{\mathrm{n}}V = \rho_{\mathrm{n}}(4\pi R^3/3)$. Thus the acceleration of gravity at the surface of the star is

$$g_{\mathrm{n}} = \frac{GM_{\mathrm{n}}}{R^2} = \frac{G\rho_{\mathrm{n}}(4\pi R^3/3)}{R^2} = \frac{4\pi}{3}G\rho_{\mathrm{n}}R\,.$$

Now consider a small chunk of mass m at the equator of the star. It is subject to two forces, $F_{\mathrm{W}} = mg_0$, which points toward the center of the circle; and F_{N}, the normal force from the surrounding masses, which points away from the center of the circle. The net force pointing towards the center of the star is then $F_{\mathrm{net}} = F_{\mathrm{W}} - F_{\mathrm{N}} = mg_0 - F_{\mathrm{N}}$. As the star rotates with a period T, the chunk undergoes circular motion with speed $v = 2\pi R/T$. The corresponding centripetal acceleration, $a_{\mathrm{c}} = v^2/R$, must be provided by F_{net}, which happens to point at the center of the circular motion (since it coincides with the center of the star). Thus Newton's Second Law for the mass reads

$$F_{\mathrm{net}} = F_{\mathrm{W}} - F_{\mathrm{N}} = mg_{\mathrm{n}} - F_{\mathrm{N}} = F_{\mathrm{c}} = m\frac{v^2}{R} = m\frac{(2\pi R/T)^2}{R} = \frac{4\pi^2 mR}{T^2}\,.$$

Obviously, T decreases as F_{N} decreases. So for minimum T we set $F_{\mathrm{N}} = 0$ to obtain $mg_{\mathrm{n}} = m(4\pi G\rho_{\mathrm{n}}R/3) = 4\pi^2 mR/T_{\mathrm{min}}^2$, which yields

$$T_{\mathrm{min}} = \sqrt{\frac{3\pi}{G\rho_{\mathrm{n}}}} = \sqrt{\frac{3\pi}{(6.67\times10^{-11}\,\mathrm{N{\cdot}m^2/kg^2})(1\times10^{17}\,\mathrm{kg/m^3})}} = 1\times10^{-3}\,\mathrm{s}\,.$$

5.67

The gravitational force F_G exerted by the Moon on the Lunar Orbiter of mass m is responsible for the centripetal force F_c necessary for the Orbiter to undergo a circular motion with speed v and radius r about the center of the Moon:

$$F_G = \frac{GmM_{\leftmoon}}{r^2} = F_c = \frac{mv^2}{r},$$

where $r = R_{\leftmoon} + 62\,\text{km} = 1738\,\text{km} + 62\,\text{km} = 1800\,\text{km}$. Solve for v:

$$v = \sqrt{\frac{GM_{\leftmoon}}{r}} = \sqrt{\frac{(6.67\times10^{-11}\,\text{m}^3/\text{kg}\cdot\text{s}^2)(7.35\times10^{22}\,\text{kg})}{1800\times10^3\,\text{m}}} = 1.65\times10^3\,\text{m/s}.$$

5.75

The gravitational force F_G exerted by the Sun on an object of mass m placed on the Earth is given by $F_G = GmM_\odot/r^2_{\odot\oplus}$. Thus

$$g_\oplus = \frac{F_G}{m} = \frac{GM_\odot}{r^2_{\odot\oplus}} = \frac{(6.67\times10^{-11}\,\text{N}\cdot\text{m}^2/\text{kg}^2)(1.987\times10^{30}\,\text{kg})}{(1.495\times10^{11}\,\text{m})^2} = 5.9\times10^{-3}\,\text{m/s}^2.$$

5.83

First, calculate the gravitational force F_G exerted on a small mass m located a distance r ($r < R$) from the center of the sphere. From the hint given in Problem (5.57) we see that the only portion of the cloud which exerts a net gravitational force F_G on the small mass is the spherical region of radius r, shown in the figure to the right as the darker-shaded portion of the cloud. Since the mass M of the cloud is uniformly distributed over its volume V, the mass M' of the shaded portion of volume V' satisfies $M'/M = V'/V = r^3/R^3$, or $M' = M(r/R)^3$, where we noted that $V' = 4\pi r^3/3 \propto r^3$ and $V \propto R^3$. Thus

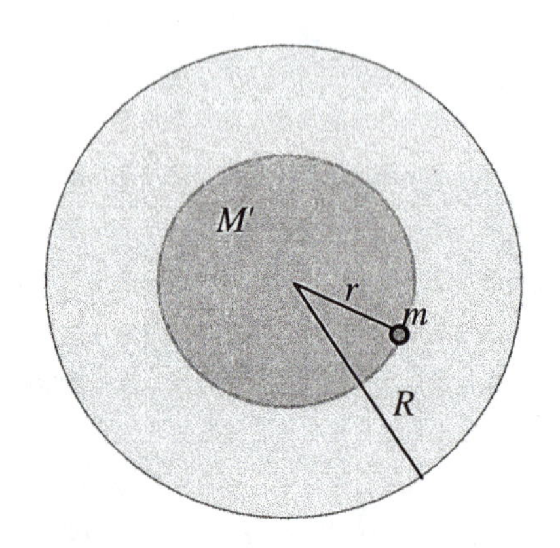

$$F_G(r) = \frac{GmM'}{r^2} = \frac{GmM(r/R)^3}{r^2} = \frac{GmMr}{R^3}.$$

Now set $r = \frac{1}{2}R$ to obtain

$$g(r) = \frac{F_G(r)}{m} = \frac{GMr}{R^3} = \frac{GM(R/2)}{R^3} = \frac{GM}{2R^2}.$$

6 *Energy*

Answers to Selected Discussion Questions

•6.1

No work is done if the rocket doesn't move. The rocket pushes down on the exhaust gas and the gas pushes up on the rocket, accelerating it. Momentum is conserved, so the momentum of the vehicle equals the momentum of the gas. The work done on the rocket increases with time as it moves faster and therefore farther per second. With respect to the ground, the gas initially gets most of the KE. As the vehicle speeds up, more energy is transferred to it. All the KE, rocket and exhaust, comes from the chemical energy of the fuel.

•6.7

Assuming both balls are launched with the same very slow speed, the ball moving along the depression wins the race. Its velocity will increase as it descends and therefore its component in the forward direction, parallel to the plane, will always be equal to or greater than the other ball's. This will continue to be the case up to some threshold launch speed. See *Phys. Teach.* **33**, 376 (1995).

•6.9

Work is energy transferred to or from a system via the application of a force acting over a distance. Historically, it got its name before the concept of energy was formalized, and so it continued to be treated as if it were something other than energy. Ergo, the notion that energy is the ability to do work. It's just as circular as saying energy is the ability to produce *vis viva* (KE). It makes sense, since work and PE both have the ability to produce *vis viva*, but it's clearly circular.

•6.11

At a constant speed there is a force, but it's perpendicular to s. When the ball has a tangential acceleration there is a tangential force — work done — and an increase (or decrease) in KE. A tangential component of the tension can exist when the string leads the ball, making an angle with $\vec{v}$ of less than 90°.

•6.13

The friction force accelerates the crate such that $W = \Delta\text{KE} = \frac{1}{2}mv^2$. The work is positive. The work overcomes inertia. The energy appears as KE. When stopping, the friction force is in the opposite direction to the displacement and does negative work, converting KE to thermal energy.

Answers to Odd-Numbered Multiple Choice Questions

1. a **3.** a **5.** a **7.** c **9.** c **11.** a **13.** a
15. c **17.** c **19.** d

Solutions to Selected Problems

6.25

The weight of the object is being supported by five segments of rope, as shown in Fig. P25. Each segment supports 1/5 of the weight. The force that is needed to pull out the rope, being equal to the tension in the rope, is therefore only 1/5 of the weight to be raised. Since work-in = work-out, in order to raise the weight by 1.0 m we need to pull out $5 \times 1.0\,\text{m} = 5.0\,\text{m}$ of the rope.

6.33

In Chapter 4 we learned that the force of friction exerted on a moving object of mass m on an incline is given by $F_f = \mu_r mg\cos\theta$, where μ_r is the coefficient of friction and θ is the angle of inclination. Over a displacement s along the inclined road, the work done by the boy in overcoming the friction is therefore

$$W = F_f s = (\mu_r mg\cos\theta)s = (0.02)(25\,\text{kg})(9.81\,\text{m/s}^2)(\cos 10°)(25\,\text{m}) = 1.2 \times 10^2\,\text{J}\,.$$

6.57

From the definition of kinetic energy in Eq. (6.13), with $m = 6.5\,\mathrm{g} = 6.5 \times 10^{-3}\,\mathrm{kg}$ and $v = 300\,\mathrm{m/s}$,

$$\mathrm{KE} = \frac{1}{2}mv^2 = \frac{1}{2}(6.5 \times 10^{-3}\,\mathrm{kg})(300\,\mathrm{m/s})^2 = 2.9 \times 10^2\,\mathrm{J} = 0.29\,\mathrm{kJ}\,.$$

6.62

Consider an object of mass m, which moves under the influence of a net force F over an infinitesimal distance ds during time dt. The work dW done by the force on the object is, by definition,

$$dW = F ds = \left(m\frac{dv}{dt}\right) ds = m\left(\frac{ds}{dt}\right) dv = mv\, dv\,,$$

where we used Newton's Second Law, $F = ma = mdv/dt$, and $v = ds/dt$. Now integrate from the initial to the final point, whereupon v changes from v_{i} to v_{f}:

$$\int_{P_{\mathrm{i}}}^{P_{\mathrm{f}}} dW = \int_{v_{\mathrm{i}}}^{v_{\mathrm{f}}} mv\, dv = m\left[\frac{1}{2}v^2\right]_{v_{\mathrm{i}}}^{v_{\mathrm{f}}} = \frac{1}{2}mv_{\mathrm{f}}^2 - \frac{1}{2}mv_{\mathrm{i}}^2\,,$$

which, with the definition $\mathrm{KE} = \frac{1}{2}mv^2$, becomes

$$W_{P_{\mathrm{i}} \to P_{\mathrm{f}}} = \mathrm{KE}_{\mathrm{f}} - \mathrm{KE}_{\mathrm{i}}\,,$$

which is the Work-Energy Theorem.

6.89

The weight of George-the-monkey is $F_{\mathrm{W}} = 10\,\mathrm{N}$, which is much less than the weight attached to the other end of the rope. So unless the monkey accelerates up the rope at an incredible rate of $a = 9g$ or greater, which is unrealistic, he will not be able to lift the 100-N weight off the floor. In the following, we will assume that he moves up the rope at a constant speed, so $a = 0$.

(a) Since we assume that the monkey climbs up the rope at a constant speed there is no change in his kinetic energy. The work he does is equal to the change in his *gravitational*-PE: $W = F_{\mathrm{W}}\Delta h = (10\,\mathrm{N})(10\,\mathrm{m}) = 1.0 \times 10^2\,\mathrm{J} = 0.10\,\mathrm{kJ}$.

(b) Since the 100-N weight is not lifted off the floor, the rope does not move, so no rope ends up on the floor.

(c) The 100-N mass does not move, so there is no change in its *gravitational*-PE. So for the system $\Delta\mathrm{PE}_{\mathrm{G}} = 0.10\,\mathrm{kJ}$ [see part (b) above)], which is due to the motion of the monkey alone.

6.105

As he jumps down from an initial height of 10.0 m to a final height of 1.00 m above the ground, the athlete's vertical displacement is $\Delta h = 10.0\,\mathrm{m} - 1.00\,\mathrm{m} = 9.00\,\mathrm{m}$. The corresponding *loss* in gravitational potential energy for the athlete (whose mass is m) is

$$-\Delta\mathrm{PE}_{\mathrm{G}} = -mg\Delta h = (55.0\,\mathrm{kg})(9.81\,\mathrm{m/s^2})(10.0\,\mathrm{m} - 1.00\,\mathrm{m}) = 4.85 \times 10^3\,\mathrm{J} = 4.85\,\mathrm{kJ}\,.$$

This is the energy stored in the trampoline.

Since there is no loss of mechanical energy, the athlete will rise back to the same initial height she started with, which is 10.0 m above the ground.

6.111

To lift a certain payload of mass m off the surface of the Earth, we must first accelerate it to at least the escape speed from the Earth: $v_{esc} = \sqrt{2GM_\oplus/R_\oplus}$ [see Eq. (6.24)]. The corresponding energy needed is

$$\mathrm{KE}_\oplus = \frac{1}{2}mv_{esc}^2 = \frac{1}{2}m\left(\frac{2GM_\oplus}{R_\oplus}\right) = \frac{GmM_\oplus}{R_\oplus}.$$

Similarly, the minimum energy needed to lift the same payload off the surface of the Moon is $\mathrm{KE}_☾ = GmM_☾/R_☾$. Take the ratio of the two energies to obtain

$$\frac{\mathrm{KE}_☾}{\mathrm{KE}_\oplus} = \frac{GmM_☾/R_☾}{GmM_\oplus/R_\oplus} = \left(\frac{M_☾}{M_\oplus}\right)\left(\frac{R_\oplus}{R_☾}\right) = \left(\frac{7.35\times10^{22}\,\mathrm{kg}}{5.975\times10^{24}\,\mathrm{kg}}\right)\left(\frac{6.371\times10^6\,\mathrm{m}}{1.74\times10^6\,\mathrm{m}}\right) = 0.0450\,,$$

which is roughly 5%.

6.113

Recall that the orbital speed v_o is given by Eq. (5.11): $v_o = \sqrt{GM_\oplus/r}$; while the escape speed satisfies Eq. (6.24): $v_{esc} = \sqrt{2GM_\oplus/r}$. (Note that, since the satellite is already at a distance r from the center of the Earth, rather than on the surface of the Earth, we must replace R with r in the expression for v_{esc}.) Compare these two expressions to obtain $v_{esc} = \sqrt{2}\,v_o$. With $v_o = 1500\,\mathrm{m/s}$, the required escape speed is

$$v_{esc} = \sqrt{2}\,v_o = \sqrt{2}\,(1500\,\mathrm{m/s}) = 2121\,\mathrm{m/s}.$$

6.131

From the problem statement of Problem (6.130) we know that the internal energy liberated by 1 liter of oxygen is 2.0×10^4 J. Since the person in question consumes oxygen at a rate of 0.40 liter/min, or $(0.40\,\mathrm{liters/min})(1\,\mathrm{min}/60\,\mathrm{s}) = 6.67\times10^{-3}\,\mathrm{liter/s}$, the rate of energy consumption for the person is $(6.67\times10^{-3}\,\mathrm{liter/s})(2.0\times10^4\,\mathrm{J/liter}) = 133\,\mathrm{W}$. Then with the definition of BMR found in Problem (6.129)

$$\mathrm{BMR} = \frac{\text{rate of energy consumption}}{\text{skin surface area}} = 133\,\mathrm{W}/1.8\,\mathrm{m} = 74\,\mathrm{W/m^2}.$$

7 *Momentum & Collisions*

Answers to Selected Discussion Questions

•7.3

When you catch the object, ordinarily your hand is free to move and bring it to rest in a relatively long time, which means that the force will be small for the same impulse and momentum change. Not so if your hand is against a table.

•7.5

The more massive ball requires the greater impulse to bring it up to speed since doing so means a greater momentum change.

Answers to Odd-Numbered Multiple Choice Questions

1. a **3.** b **5.** d **7.** b **9.** e **11.** c **13.** a
15. c **17.** a **19.** d

Solutions to Selected Problems

7.7

The first force F_1 is along the positive x-direction and delivers an impulse $F_1 \Delta t_1 = (+20\,\mathrm{N}) \times (2.0\,\mathrm{s}) = +40\,\mathrm{N{\cdot}s}$; while the second force F_2 is along the negative x-direction, delivering an impulse $F_2 \Delta t_2 = (-2.0\,\mathrm{N})(20\,\mathrm{s}) = -40\,\mathrm{N{\cdot}s}$. Both impulses are represented by the corresponding shaded areas in the force vs time plot shown below, with the area below the t-axis counted as negative. The net impulse experienced by the body is then

$$F_1 \Delta t_1 + F_2 \Delta t_2 = +40\,\mathrm{N{\cdot}s} - 40\,\mathrm{N{\cdot}s} = 0\,;$$

so there is no net change in the momentum of the body, whose final momentum p_f must then be the same as its initial value:

$$p_\mathrm{f} = p_\mathrm{i} = m v_\mathrm{i} = (1.0\,\mathrm{kg})(10\,\mathrm{m/s}) = 10\,\mathrm{kg{\cdot}m/s}\,.$$

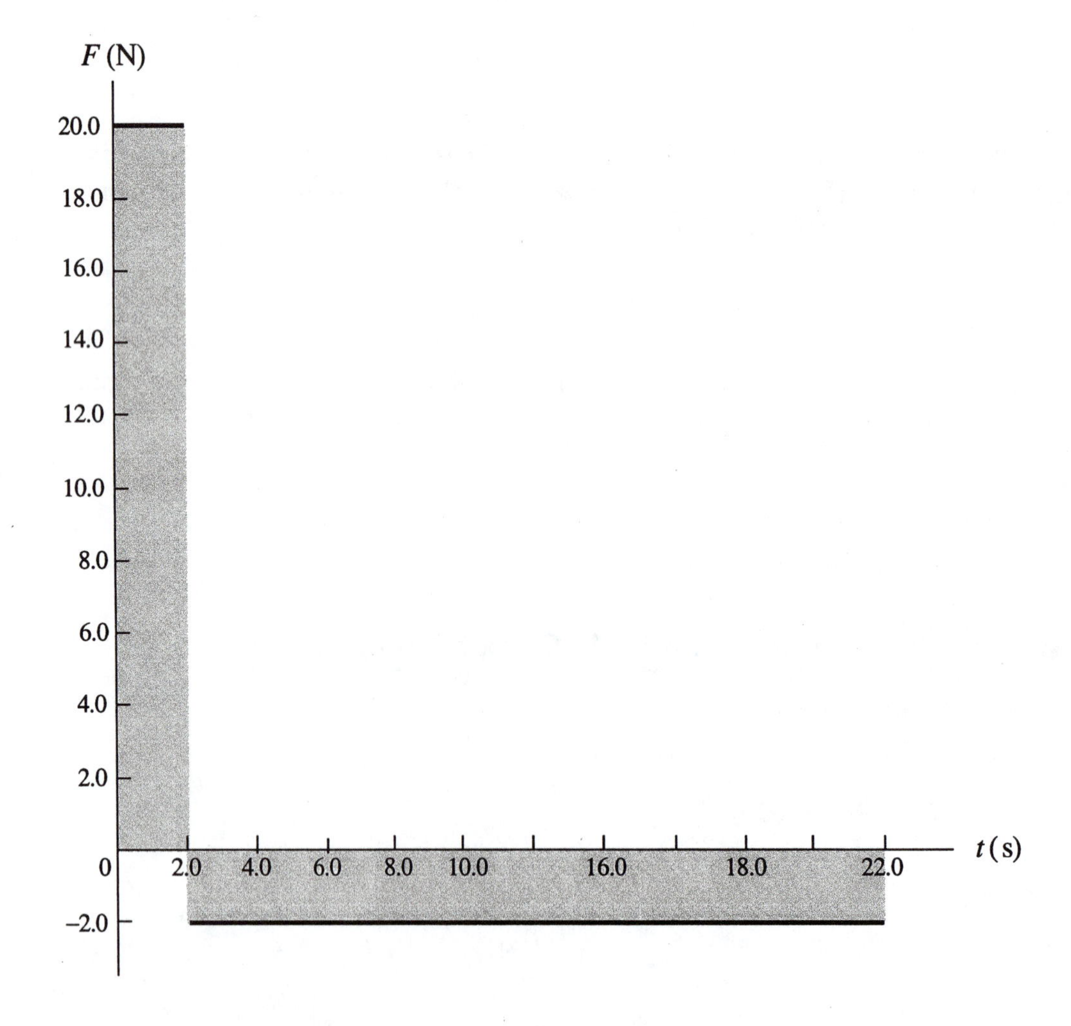

7.15

The impulse delivered on the cart by the force, $F(t) = (15.0\,\mathrm{N/s})t$, over the time interval between $t_\mathrm{i} = 0$ and $t_\mathrm{f} = 6.00\,\mathrm{s}$ is given by

$$\int_{t_\mathrm{i}}^{t_\mathrm{f}} F(t)\,dt = \int_0^{6.00\,\mathrm{s}} (15.0\,\mathrm{N/s})t\,dt = (15.0\,\mathrm{N/s})\left[\frac{1}{2}t^2\right]_0^{6.00\,\mathrm{s}} = 270\,\mathrm{N{\cdot}s}\,.$$

The resulting change in momentum for the cart is $\Delta p = p - p_\mathrm{i} = mv - mv_\mathrm{i}$, where $m = 1.00\,\mathrm{kg}$ is its mass, v its speed at $t = 6.00\,\mathrm{s}$, and $v_\mathrm{i} = 0$ its initial speed at $t = 0$. According to Eq. (7.5) the change in the cart's momentum should equal the impulse delivered on it, i.e., $\Delta p = mv - mv_\mathrm{i} = 270\,\mathrm{N{\cdot}s}$. Solve for v:

$$v = v_\mathrm{i} + \frac{270\,\mathrm{N{\cdot}s}}{m} = 0 + \frac{270\,\mathrm{N{\cdot}s}}{1.00\,\mathrm{kg}} = 270\,\mathrm{m/s}\,.$$

7.43

The initial momentum of the system consisting the cosmonaut (C) and the spaceship (S) was $\vec{\mathbf{p}}_\mathrm{i} = 0$, since neither was moving. As the cosmonaut sails towards her copilot with a velocity $\vec{\mathbf{v}}_\mathrm{C}$, her momentum is $m_\mathrm{C}\vec{\mathbf{v}}_\mathrm{C}$. Meanwhile, the spaceship is moving backward at a velocity $\vec{\mathbf{v}}_\mathrm{S}$, resulting in a momentum of $m_\mathrm{S}\vec{\mathbf{v}}_\mathrm{S}$. The total momentum of the system is now $\vec{\mathbf{p}}_\mathrm{f} = m_\mathrm{C}\vec{\mathbf{v}}_\mathrm{C} + m_\mathrm{S}\vec{\mathbf{v}}_\mathrm{S}$. Conservation of momentum requires that $\vec{\mathbf{p}}_\mathrm{i} = \vec{\mathbf{p}}_\mathrm{f}$, which becomes

$$p_\mathrm{i} = 0 = p_\mathrm{f} = m_\mathrm{C}v_\mathrm{C} + m_\mathrm{S}v_\mathrm{S}$$

in scalar form. Take the direction of $\vec{\mathbf{v}}_\mathrm{C}$ as positive and solve for v_S, the velocity of the spaceship:

$$v_\mathrm{S} = -\frac{m_\mathrm{C}v_\mathrm{C}}{m_\mathrm{S}} = -\frac{(60\,\mathrm{kg})(10\,\mathrm{m/s})}{5000\,\mathrm{kg}} = -0.12\,\mathrm{m/s}\,,$$

where the minus sign indicates that $\vec{\mathbf{v}}_\mathrm{S}$ is opposite in direction to $\vec{\mathbf{v}}_\mathrm{C}$.

7.49

Since friction is negligible, the horizontal momentum of the skater-snowball system is conserved. The initial momentum of the system is $\vec{\mathbf{p}}_\mathrm{i} = 0$, since neither the skater (S) nor the snowball (B) was moving. As the snowball is thrown out at a velocity $\vec{\mathbf{v}}_\mathrm{B}$, whose magnitude is $v_\mathrm{B} = (20.0\,\mathrm{km/h})(10^3\,\mathrm{m/km})(1\,\mathrm{h}/3600\,\mathrm{s}) = 5.556\,\mathrm{m/s}$, it picks up a momentum $m_\mathrm{B}\vec{\mathbf{v}}_\mathrm{B}$. Meanwhile, the skater is moving backward at a velocity $\vec{\mathbf{v}}_\mathrm{S}$, resulting in a momentum of $m_\mathrm{S}\vec{\mathbf{v}}_\mathrm{S}$. The total momentum of the system is now $\vec{\mathbf{p}}_\mathrm{f} = m_\mathrm{B}\vec{\mathbf{v}}_\mathrm{B} + m_\mathrm{S}\vec{\mathbf{v}}_\mathrm{S}$. From conservation of momentum $\vec{\mathbf{p}}_\mathrm{i} = \vec{\mathbf{p}}_\mathrm{f}$, which becomes

$$p_\mathrm{i} = 0 = p_\mathrm{f} = m_\mathrm{B}v_\mathrm{B} + m_\mathrm{S}v_\mathrm{S}$$

in scalar form. Take the direction of $\vec{\mathbf{v}}_\mathrm{B}$ as positive and solve for v_S, the velocity of the skater:

$$v_\mathrm{S} = -\frac{m_\mathrm{B}v_\mathrm{B}}{m_\mathrm{S}} = -\frac{(200\times10^{-3}\,\mathrm{kg})(5.556\,\mathrm{m/s})}{55.0\,\mathrm{kg}} = -0.020\,2\,\mathrm{m/s}\,,$$

where the minus sign indicates that $\vec{\mathbf{v}}_\mathrm{S}$ is opposite in direction to $\vec{\mathbf{v}}_\mathrm{B}$.

7.75

Since the angle ϕ between the final velocities of the two balls, labeled 1 and 2, respectively, is independent of the reference frame in which it is measured, we may, without loss of generality, switch to a new reference frame in which only one of the balls, say ball 1, is moving prior to the collision. The initial momentum of the two-ball system is then $\vec{\mathbf{p}}_i = m\vec{\mathbf{v}}_{1i}$. After the collision the final momentum of the system is $\vec{\mathbf{p}}_f = m\vec{\mathbf{v}}_{1f} + m\vec{\mathbf{v}}_{2f}$. Here m is the mass of each ball. Conservation of momentum then leads to $m\vec{\mathbf{v}}_{1i} = m\vec{\mathbf{v}}_{1f} + m\vec{\mathbf{v}}_{2f}$, or

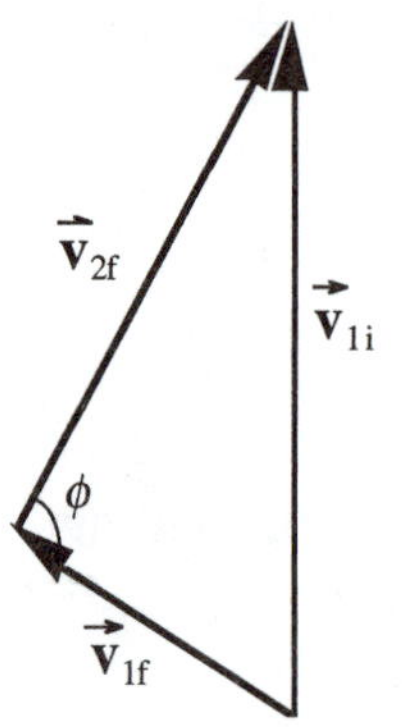

$$\vec{\mathbf{v}}_{1i} = \vec{\mathbf{v}}_{1f} + \vec{\mathbf{v}}_{2f} .$$

$\vec{\mathbf{v}}_{2f}$ $\vec{\mathbf{v}}_{1i}$ ϕ $\vec{\mathbf{v}}_{1f}$

The resulting vector triangle, consisting of $\vec{\mathbf{v}}_{1i}$, $\vec{\mathbf{v}}_{1f}$ and $\vec{\mathbf{v}}_{2f}$, are shown above on the right. Also, conservation of kinetic energy requires that $\text{KE}_i = \frac{1}{2}mv_{1i}^2 = \text{KE}_f = \frac{1}{2}mv_{1f}^2 + \frac{1}{2}mv_{2f}^2$, or

$$v_{1i}^2 = v_{1f}^2 + v_{2f}^2 ,$$

which means that the three sides of this vector triangle satisfy the Pythagorean theorem. So the triangle must be a right-angled one, with $\vec{\mathbf{v}}_{1i}$ as the hypotenuse. The other two sides, $\vec{\mathbf{v}}_{1f}$ and $\vec{\mathbf{v}}_{2f}$, must then be perpendicular to each other (i.e., $\phi = 90°$ in the figure above).

8 Rotational Motion

Answers to Selected Discussion Questions

•8.1
The arm pivots on the end of the humerus. To raise a load, the biceps contracts, while the triceps stays relaxed; to push down, the triceps contracts, while the biceps relaxes.

•8.5
The outside wheels turn faster. A differential gear is mounted between pairs of wheels.

•8.11
It must be zero since the angular momentum does not change, and $\vec{\tau} = 0$.

•8.15
L in the ball-person system is not conserved because there's a friction force via the ground on the feet. Enlarging the system to include the Earth conserves angular momentum, and so the Earth's motion must change accordingly. The person would both rotate and translate.

•8.19
At a constant speed, the forward propeller reaction force $\vec{\mathbf{F}}$, which is beneath the *c.m.*, equals the friction force arising from the air and water, just as the weight equals the upward buoyant force, and so $\sum \vec{\mathbf{F}} = 0$. At any moment (neglecting friction and any vertical component from the propeller force), there is a balance between the torque from the propeller force and the buoyant force. The greater $\vec{\mathbf{F}}$ is made, the more the boat noses up and the farther toward the

rear (away from the *c.m.*) the point of action of the buoyant force moves, thereby increasing its torque until equilibrium is reached again.

•8.21
Nothing.

Answers to Odd-Numbered Multiple Choice Questions

1. c	**3.** a	**5.** e	**7.** d	**9.** e	**11.** b	**13.** c
15. b	**17.** b	**19.** a	**21.** d	**23.** b		

Solutions to Selected Problems

8.11
The standard TV picture has 525 horizontal scan lines so the separation between two adjacent lines is $\ell = 30\,\text{cm}/525 = 5.71 \times 10^{-4}\,\text{m}$ for a screen 30 cm high. The angle θ that any pair of adjacent scan lines subtend where you are is then $\theta = \ell/r$, where r is the distance between you and the TV screen. If you can no longer distinguish such two adjacent lines, then θ must be no more than $\theta_{\text{min}} = (1.0\,\text{min})(1°/60\,\text{min})(\pi\,\text{rad}/180°) = 2.91 \times 10^{-4}\,\text{rad}$, which is the smallest angle your eyes can resolve. Let $\theta_{\text{min}} = \ell/r_{\text{min}} \le \theta = \ell/r$, we may find r_{min}, the minimum distance you should sit in front of the TV:

$$r_{\text{min}} = \frac{\ell}{\theta_{\text{min}}} = \frac{5.71 \times 10^{-4}\,\text{m}}{2.91 \times 10^{-4}\,\text{rad}} = 2.0\,\text{m}\,.$$

8.37
The angular speed ω of the train moving in a circle of radius r is related to its linear speed v via Eq. (8.10): $v = r\omega$. Solve for ω:

$$\omega = \frac{v}{r} = \frac{8.9\,\text{m/s}}{304.8\,\text{m}} = 0.029\,\text{rad/s}\,.$$

The centripetal acceleration of the train is

$$a_c = \frac{v^2}{r} = r\omega^2 = (304.8\,\text{m})(0.029\,2\,\text{rad/s})^2 = 0.26\,\text{m/s}^2\,.$$

8.39

The linear speed at the perimeter of the pulley on the left (pulley 1) is the same as that at the perimeter of the inner pulley on the right (pulley 2): $v_1 = \omega_1 R_1 = v_2 = \omega_2 R_2$. The linear speed of suspended body, v_B, is the same as the linear speed v_3 at the perimeter of the larger pulley (pulley 3) on the right: $v_B = v_3 = \omega_2 R_3$. Here $\omega_1 = 1.0\,\text{rpm} = (1.0\,\text{rev/min})(2\pi\,\text{rad/rev}) \times (1\,\text{min}/60\,\text{s}) = 0.105\,\text{rad/s}$. Solve for ω_2 from the first equation above and plug the result into the second equation for v_B:

$$v_B = \omega_2 R_3 = \left(\frac{\omega_1 R_1}{R_2}\right) R_3 = \frac{(0.105\,\text{rad/s})(0.3\,\text{m})(0.4\,\text{m})}{0.1\,\text{m}} = 0.1\,\text{m/s}\,.$$

Since the pulleys on the right turn counterclockwisely as the driver turns clockwisely, the suspended body is being pulled upward.

8.67

The pulley (P) and turntable platter (T) are linked via the same belt so they must have the same linear speed: $v_P = R_P\omega_P = v_T = R_T\omega_T$. This gives $\omega_P = \omega_T(R_T/R_P)$. Since $\omega_P = \alpha_P t$ and $\omega_T = \alpha_T t$, the equation above relating ω_T and ω_P leads to

$$\alpha_P = \alpha_T\left(\frac{R_T}{R_P}\right)\,.$$

Noting that $\alpha_T = \omega_T/t$, where $\omega_T = (33\frac{1}{3}\,\text{rev/min})(2\pi\,\text{rad/rev})(1\,\text{min}/60\,\text{s}) = 3.49\,\text{rad/s}$ and $t = 6.0\,\text{s}$, we obtain

$$\alpha_P = \alpha_T\left(\frac{R_T}{R_P}\right) = \left(\frac{3.49\,\text{rad/s}}{6.0\,\text{s}}\right)\left(\frac{15\,\text{cm}}{1.0\,\text{cm}}\right) = 8.7\,\text{rad/s}^2\,.$$

From Eq. (8.24) we then obtain the angle θ_P through which the pulley turns in the process:

$$\theta_P = \omega_{P0}t + \frac{1}{2}\alpha_P t^2 = \frac{1}{2}(8.7\,\text{rad/s}^2)(6.0\,\text{s})^2 = 156.6\,\text{rad}\,,$$

which amounts to 25 rev. Here we noted that ω_{P0}, the initial angular speed of the pulley, is zero.

8.71

Use Eq. (8.26) to find the torque of a force F about point O: $\tau_0 = Fr_\perp$. Here $F = 100\,\text{N}$ and $r_\perp = 1.00\,\text{m}$ (which is the distance between the pivot point O and the line-of-action of the 100-N force); so

$$\tau_0 = Fr_\perp = (100\,\text{N})(1.00\,\text{m}) = 100\,\text{N·m}\,,$$

clockwise.

8.79

Consider a vector $\vec{\mathbf{r}} = (2.00\,\mathrm{m})\,\hat{\mathbf{i}}$ that points from the center of the wheel to a point on its rim. Since the force $\vec{\mathbf{F}}$ passes through the end of $\vec{\mathbf{r}}$, from Eq. (8.30)

$$\vec{\tau} = \vec{\mathbf{r}} \times \vec{\mathbf{F}} = (2.00\,\mathrm{m})\,\hat{\mathbf{i}} \times (4.50\,\mathrm{N})\,\hat{\mathbf{j}} = (2.00\,\mathrm{m})(4.50\,\mathrm{N})(\hat{\mathbf{i}} \times \hat{\mathbf{j}}) = (9.00\,\mathrm{N{\cdot}m})\,\hat{\mathbf{k}}\,.$$

8.81

From Eq. (8.30)

$$\begin{aligned}
\vec{\tau} &= \vec{\mathbf{r}} \times \vec{\mathbf{F}} \\
&= \Big[(2.00\,\mathrm{m})\,\hat{\mathbf{i}} + (1.00\,\mathrm{m})\,\hat{\mathbf{j}}\Big] \times \Big[(10.5\,\mathrm{N})\,\hat{\mathbf{i}} + (4.00\,\mathrm{N})\,\hat{\mathbf{j}} + (2.50\,\mathrm{N})\,\hat{\mathbf{k}}\Big] \\
&= (2.00\,\mathrm{m})(10.5\,\mathrm{N})(\hat{\mathbf{i}} \times \hat{\mathbf{i}}) + (2.00\,\mathrm{m})(4.00\,\mathrm{N})(\hat{\mathbf{i}} \times \hat{\mathbf{j}}) + (2.00\,\mathrm{m})(2.50\,\mathrm{N})(\hat{\mathbf{i}} \times \hat{\mathbf{k}}) + \\
&\quad (1.00\,\mathrm{m})(10.5\,\mathrm{N})(\hat{\mathbf{j}} \times \hat{\mathbf{i}}) + (1.00\,\mathrm{m})(4.00\,\mathrm{N})(\hat{\mathbf{j}} \times \hat{\mathbf{j}}) + (1.00\,\mathrm{m})(2.50\,\mathrm{N})(\hat{\mathbf{j}} \times \hat{\mathbf{k}}) \\
&= 0 + (8.00\,\mathrm{N{\cdot}m})\,\hat{\mathbf{k}} + (5.00\,\mathrm{N{\cdot}m})(-\hat{\mathbf{j}}) + (10.5\,\mathrm{N{\cdot}m})(-\hat{\mathbf{k}}) + 0 + (2.50\,\mathrm{N{\cdot}m})\,\hat{\mathbf{i}} \\
&= (2.50\,\mathrm{N{\cdot}m})\,\hat{\mathbf{i}} - (5.00\,\mathrm{N{\cdot}m})\,\hat{\mathbf{j}} - (2.5\,\mathrm{N{\cdot}m})\,\hat{\mathbf{k}}\,.
\end{aligned}$$

8.82

From Eq. (8.30)

$$\begin{aligned}
\vec{\tau} &= \vec{\mathbf{r}} \times \vec{\mathbf{F}} \\
&= \Big[(2.00\,\mathrm{m})\,\hat{\mathbf{i}} + (1.00\,\mathrm{m})\,\hat{\mathbf{j}}\Big] \times (2.50\,\mathrm{N})\,\hat{\mathbf{k}} \\
&= (2.00\,\mathrm{m})(2.50\,\mathrm{N})(\hat{\mathbf{i}} \times \hat{\mathbf{k}}) + (1.00\,\mathrm{m})(2.50\,\mathrm{N})(\hat{\mathbf{j}} \times \hat{\mathbf{k}}) \\
&= (5.00\,\mathrm{N{\cdot}m})(-\hat{\mathbf{j}}) + (2.50\,\mathrm{N{\cdot}m})\,\hat{\mathbf{i}} \\
&= (2.50\,\mathrm{N{\cdot}m})\,\hat{\mathbf{i}} - (5.00\,\mathrm{N{\cdot}m})\,\hat{\mathbf{j}}\,.
\end{aligned}$$

8.83

For the beam to be in mechanical equilibrium the net torque exerted on it about the pivot point (denoted as O) must be zero. Let $F_{\mathrm{W1}} = (2.0\,\mathrm{kg})g$ be the weight of the hanging mass on the left, $F_{\mathrm{W2}} = (1.0\,\mathrm{kg})g$ be that of the one on the right, and $l = 10\,\mathrm{cm}$, then the reaction force F_{RB} of the scale satisfies

$$\overset{+}{\circlearrowleft}\sum \tau_{\mathrm{o}} = -F_{\mathrm{W1}}\,l - F_{\mathrm{RB}}\,(2l) + F_{\mathrm{W2}}\,(4l) = 0\,,$$

which gives

$$F_{\mathrm{RB}} = 2F_{\mathrm{W2}} - \frac{1}{2}F_{\mathrm{W1}} = 2(1.0\,\mathrm{kg})g - \frac{1}{2}(2.0\,\mathrm{kg})g = (1.0\,\mathrm{kg})g\,.$$

Thus the reading of the scale, in kilograms, is $F_{RB}/g = 1.0\,\text{kg}$.

8.89

The displacement vector representing the position of the dot is $\vec{\mathbf{r}} = (2.00\,\text{m})\,\hat{\mathbf{i}} + (3.00\,\text{m})\,\hat{\mathbf{j}} + (1.00\,\text{m})\,\hat{\mathbf{k}}$. Thus from Eq. (8.30)

$$\begin{aligned}\vec{\tau} &= \vec{\mathbf{r}} \times \vec{\mathbf{F}} \\ &= \left[(2.00\,\text{m})\,\hat{\mathbf{i}} + (3.00\,\text{m})\,\hat{\mathbf{j}} + (1.00\,\text{m})\,\hat{\mathbf{k}}\right] \times \left[(4.50\,\text{N})\,\hat{\mathbf{i}} + (2.00\,\text{N})\,\hat{\mathbf{j}}\right] \\ &= (2.00\,\text{m})(4.50\,\text{N})(\hat{\mathbf{i}} \times \hat{\mathbf{i}}) + (2.00\,\text{m})(2.00\,\text{N})(\hat{\mathbf{i}} \times \hat{\mathbf{j}}) + (3.00\,\text{m})(4.50\,\text{N})(\hat{\mathbf{j}} \times \hat{\mathbf{i}}) + \\ &\quad (3.00\,\text{m})(2.00\,\text{N})(\hat{\mathbf{j}} \times \hat{\mathbf{j}}) + (1.00\,\text{m})(4.50\,\text{N})(\hat{\mathbf{k}} \times \hat{\mathbf{i}}) + (1.00\,\text{m})(2.00\,\text{N})(\hat{\mathbf{k}} \times \hat{\mathbf{j}}) \\ &= 0 + (4.00\,\text{N}\cdot\text{m})\,\hat{\mathbf{k}} + (13.5\,\text{N}\cdot\text{m})(-\hat{\mathbf{k}}) + 0 + (4.50\,\text{N}\cdot\text{m})\,\hat{\mathbf{j}} + (2.00\,\text{N}\cdot\text{m})(-\hat{\mathbf{i}}) \\ &= -(2.00\,\text{N}\cdot\text{m})\,\hat{\mathbf{i}} + (4.50\,\text{N}\cdot\text{m})\,\hat{\mathbf{j}} - (9.5\,\text{N}\cdot\text{m})\,\hat{\mathbf{k}}\,.\end{aligned}$$

8.90

Consider the torques due to the three forces with respect to the left end of the rod. Obviously the torque $\vec{\tau}_1$ due to $\vec{\mathbf{F}}_1$ is zero, since $\vec{\mathbf{F}}_1$ passes right through that point. The torque due to $\vec{\mathbf{F}}_2$ is

$$\vec{\tau}_2 = \vec{\mathbf{r}}_2 \times \vec{\mathbf{F}}_2 = (2.0\,\text{m})\,\hat{\mathbf{i}} \times (100\,\text{N})\,\hat{\mathbf{j}} = (0.20\,\text{kN}\cdot\text{m})(\hat{\mathbf{i}} \times \hat{\mathbf{j}}) = (0.20\,\text{kN}\cdot\text{m})\,\hat{\mathbf{k}}\,,$$

where $\vec{\mathbf{r}}_2$ is a vector starting from the left end of the rod and ending at the line-of-action of $\vec{\mathbf{F}}_2$. Similarly, if $\vec{\mathbf{F}}_3$ is applied a distance x from the left end of the rod then its torque is

$$\vec{\tau}_3 = \vec{\mathbf{r}}_3 \times \vec{\mathbf{F}}_3 = (x\,\hat{\mathbf{i}}) \times \left[(-100\,\text{N})\,\hat{\mathbf{i}} - (50\,\text{N})\,\hat{\mathbf{j}}\right] = -(50\,\text{N})x\,\hat{\mathbf{k}}\,.$$

For the rod to be in equilibrium

$$\sum \vec{\tau} = \vec{\tau}_1 + \vec{\tau}_2 + \vec{\tau}_3 = 0 + (0.20\,\text{kN}\cdot\text{m})\,\hat{\mathbf{k}} - (50\,\text{N})x\,\hat{\mathbf{k}} = 0\,,$$

which we solve for x: $x = 0.20\,\text{kN}\cdot\text{m}/50\,\text{N} = 4.0\,\text{m}$.

8.99

Since the suspended body is in mechanical equilibrium, the three forces exerted on it must sum up to zero: $\sum \vec{\mathbf{F}} = \vec{\mathbf{F}}_{T1} + \vec{\mathbf{F}}_{T2} + \vec{\mathbf{F}}_{W} = 0$. Here $\vec{\mathbf{F}}_{T1}$ is the force exerted by Scale-1, $\vec{\mathbf{F}}_{T2}$ is that exerted by Scale-2, and $\vec{\mathbf{F}}_{W}$ is the weight of the body. The horizontal component of the vector equation above reads

$$\overset{+}{\rightarrow}\sum F_x = F_{T2}\cos 60.0^\circ - F_{T1}\cos 60.0^\circ = 0\,,$$

which gives $F_{T2} = F_{T1} = 100\,\text{N}$. To find the weight of the body, write down the vertical component of the sum-of-force equation:

$$+\uparrow\sum F_y = F_{T1}\sin 60.0^\circ + F_{T2}\sin 60.0^\circ - F_{W} = 0\,,$$

so $F_{W} = F_{T1}\sin 60.0^\circ + F_{T2}\sin 60.0^\circ = 2(100\,\text{N})(\sin 60.0^\circ) = 173\,\text{N}$.

The line-of-action of each of the two tension forces intersect at a point directly below the *c.g.* This is because $\vec{\mathbf{F}}_W$ must pass through the same point to make the net torque on the body zero.

8.109

The net torque and force exerted on the woman must both vanish for her to be in balance. Let the forces exerted on her hands (H) and feet (F) be F_H and F_F, respectively. Then, measured about her *c.g.*,

$$\overset{+}{\circlearrowleft}\sum \tau = F_H(0.50\,\text{m}) - F_F(1.00\,\text{m}) = 0$$

and

$$+\uparrow\sum F_y = F_H + F_F - F_W = 0\,.$$

Plug in $F_W = (65\,\text{kg})(9.81\,\text{m/s}^2) = 638\,\text{N}$ and solve for F_H and F_F to obtain $F_H = 0.42\,\text{kN}$ and $F_F = 0.22\,\text{kN}$. This means that the force on each hand is 0.21 kN, and that on each foot is 0.11 kN.

8.135

Apply Newton's Second Law to the linear motion of the block:

$$+\downarrow\sum F_y = F_W - F_T = ma\,,$$

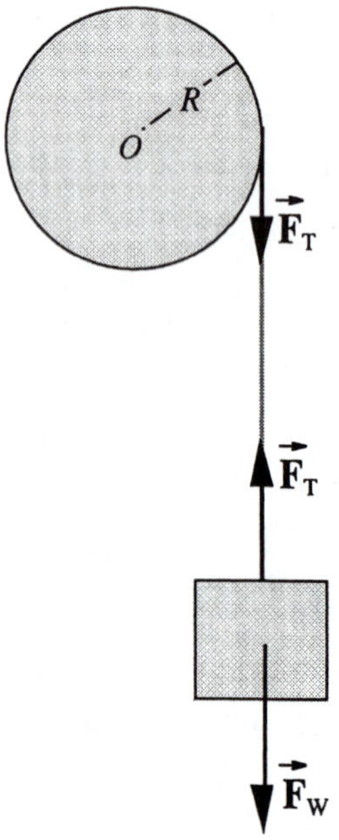

where $F_W = mg$, F_T is the tension in the cord, and a is the downward acceleration of the block. For the rotation of the pulley about point O, its center-of-mass, the only torque is that due to the tension F_T in the pulley, with a lever arm of $r_\perp = R$. So

$$\overset{+}{\circlearrowleft}\sum \tau_0 = F_T R = I\alpha\,.$$

Also, if the cord does not slip on the pulley,

$$a = \alpha R\,.$$

Solve these equations for a, the acceleration of the block:

$$a = \frac{mg}{m + I/R^2}\,.$$

8.143

Divide the rod of length l and mass M into 100 equal parts. The mass of each part is $\Delta M = M/100$. Label the one closest to the pivot point O with subscript 0, the next with subscript 1, etc. Then the separation between the center of the i-th part and the pivot point is $r_i = l/200 + (l/100)i$, and the contribution of that part to the moment-of-inertia is approximately

$\Delta I_i \approx (M/100)r_i^2 = (M/100)(l/200 + il/100)^2$. Summing over the contribution of all the parts, we get

$$I = \sum_{i=1}^{100} \Delta I_i \approx \sum_{i=0}^{99} \left(\frac{M}{100}\right)\left(\frac{l}{200} + \frac{il}{100}\right)^2 = \frac{Ml^2}{100(200)^2} \sum_{i=0}^{99} (2i+1)^2 .$$

The sum in the last step above can be evaluated using the formula given in the problem statement, with $(2n-1) = 2i_{max} + 1 = 2 \times 99 + 1$, or $n = 100$. Thus

$$\begin{aligned}\sum_{i=0}^{99} (2i+1)^2 &= 1^2 + 3^2 + 5^2 + \cdots + (2 \times 100 - 1)^2 \\ &= \frac{100}{3}(2 \times 100 + 1)(2 \times 100 - 1) \\ &= \frac{1}{3}(100)(201)(199) ,\end{aligned}$$

and so

$$I \approx \frac{1}{3}(100)(201)(199)\left[\frac{Ml^2}{100(200)^2}\right] = \frac{(201)(199)}{(200)^2}\left(\frac{1}{3}Ml^2\right) .$$

Here the factor $(201)(199)/200^2 = 0.999\,975$ is very close to 1. In fact if we choose to divide the rod into an infinite number of equal parts (which is of course more accurate), the factor will become exactly 1. So finally

$$I = \frac{1}{3}Ml^2 .$$

9 *Solids, Liquids, & Gases*

Answers to Selected Discussion Questions

•9.1

Aluminum — foil, pots; tungsten — light bulbs, in steel knife blades; carbon — in wood, pencil lead, diamond; mercury — in fluorescent bulbs, silent switches, thermometers; zinc — paint pigment, ointments, in brass; chlorine — bleach, drinking water, salt; copper — pots, wire, jewelry, in brass and bronze; sodium — table salt, Alka-Seltzer; iron — blood, raisins, nails, pots; lead — plumbing, solder, old paint; magnesium — griddles, antacid pills, sparklers; cobalt — blue dyes and pigments; oxygen — air, water, rust; chromium — in steel, plated on toasters and car bumpers; nickel — alloyed in coins, in common Alnico magnets.

•9.3

(a) Increase as the effective g increases. (b) Zero.

•9.7

(a) Nothing — you now displace one-cup's weight more of water, so the level stays the same. (b) The bust floats partially submerged, displacing its weight of water. When it was aboard, it was displacing its weight as well, so the level again does not change. (c) Nothing. (d) The level drops.

•9.11

$P_l = P_T + \rho gh$ hence $P_T = P_l - \rho gh$ and when $P_l < 0$, $P_T < 0$.

•9.13
The soap drops the surface tension in the middle and the surface pulls out to the periphery, as if you had punched a hole in a stretched rubber sheet.

•9.15
Air streaming over it asymmetrically (the ball drops a little, allowing most of the air to rush over it) produces a pressure drop above the ball (where a lot of air is moving rapidly), resulting in lift.

•9.17
Air beneath the ball will move more rapidly than above; a pressure drop below the ball will cause it to sink more swiftly than if it was not spinning.

•9.19
To have lift, the wing must have air circulating around it, which means angular momentum. Somehow, an equal and opposite amount of angular momentum will be imparted to the air-plane system and that's done nicely with the creation of a starting vortex revolving in the opposite direction. If the wing is stopped, the circulating air will generate another vortex equal and opposite to the starting vortex.

•9.21
Work is done filling the balloon and displacing the atmosphere. *Gravitational*-PE is stored in the balloon-atmosphere system. As the balloon ascends, the atmosphere descends.

Answers to Odd-Numbered Multiple Choice Questions

1. d **3.** d **5.** b **7.** a **9.** a **11.** a **13.** d
15. b **17.** c **19.** c **21.** d

Solutions to Selected Problems

9.3

Use Eq. (9.1) to find the density ρ:

$$\rho = \frac{m}{V} = \frac{(2000\,\text{lb})(0.454\,\text{kg/lb})}{[(1\,\text{in.})(2.54 \times 10^{-2}\,\text{m/in.})]^3} = 6 \times 10^7\,\text{kg/m}^3\,.$$

9.17

The density of water is $\rho = 1.000 \times 10^3\,\text{kg/m}^3$ and its molar mass is $M_m = 0.018\,\text{kg/mol}$. So from the result of the previous problem

$$\begin{aligned} N &= \frac{\rho N_A}{M_m} \\ &= \frac{(1.000 \times 10^3\,\text{kg/m}^3)(6.02 \times 10^{23}\,\text{molecules/mol})}{0.018\,\text{kg/mol}} \\ &= 3.3 \times 10^{28}\,\text{molecules/m}^3\,. \end{aligned}$$

9.21

The size of an ethyl alcohol molecule can be estimated from Eq. (9.2): $L = (M_m/N_A\rho)^{1/3}$. Since each C_2H_5OH molecule contains 2 carbon atoms, 6 hydrogen atoms, and 1 oxygen atom, its molecular mass is $2 \times 12\,\text{u} + 6 \times 1\,\text{u} + 1 \times 16\,\text{u} = 46\,\text{u}$, meaning that its molar mass is $M_m = 46 \times 10^{-3}\,\text{kg/mol}$. Plug this result into Eq. (9.2) to obtain

$$\begin{aligned} L &= \left(\frac{M_m}{N_A\rho}\right)^{1/3} = \left[\frac{46 \times 10^{-3}\,\text{kg/mol}}{(6.02 \times 10^{23}/\text{mol})(0.789 \times 10^3\,\text{kg/m}^3)}\right]^{1/3} \\ &= 4.6 \times 10^{-10}\,\text{m} = 0.46\,\text{nm}\,. \end{aligned}$$

9.31

$$\frac{1.000\,\text{lb}}{\text{in.}^2} = \frac{(1.000\,\text{lb})(4.448\,23\,\text{N/lb})}{[(1.000\,\text{in.})(0.025\,40\,\text{m/in.})]^2} = 6.895 \times 10^3\,\text{N/m}^2\,.$$

9.39

According to Eq. (9.4) the pressure due to the weight of the liquid of constant density ρ is $P_l = \rho g z$ at a depth z below the surface of the liquid. Taking into consideration the atmospheric pressure P_s at the surface of the liquid, the total pressure P at depth z is

$$P = P_s + P_l = P_s + \rho g z\,.$$

The rate-of-change of pressure with depth is then

$$\frac{dP}{dz} = \frac{d}{dz}(P_s + \rho g z) = \rho g \,,$$

where we noted that P_s, ρ and g are constants that are independent of z.

9.47

From the problem statement we know that P_G, the lowest gauge pressure most people can create in their lungs, is equal to that due to a water column of height $h = 1.1\,\text{m}$. Thus

$$P_G = -\rho_W g h = -(1.00 \times 10^3\,\text{kg/m}^3)(9.81\,\text{m/s}^2)(1.1\,\text{m}) = -1.1 \times 10^4\,\text{Pa}\,.$$

Here the minus sign indicates that the pressure inside the lung is lower than that outside.

9.85

Consider an iceberg of volume V, whose weight is given by $F_W = \rho_{\text{ice}} g V$. If the volume of the submerged portion of the iceberg is V_f, then according to Eq. (9.9) the buoyant force exerted on it by the sea water (of density ρ_f) is $F_B = \rho_f g V_f$. For the iceberg to be in mechanical equilibrium F_W and F_B must be equal: $F_W = \rho_{\text{ice}} g V = F_B = \rho_f g V_f$, which gives $V_f/V = \rho_{\text{ice}}/\rho_f$. The fraction of the volume of the iceberg submerged is V_f/V; hence the fraction of its volume that is visible (i.e. above the sea level) is given by

$$1 - \frac{V_f}{V} = 1 - \frac{\rho_{\text{ice}}}{\rho_f} = 1 - \frac{0.92 \times 10^3\,\text{kg/m}^3}{1.025 \times 10^3\,\text{kg/m}^3} = 0.10 = 10\%\,.$$

For fresh water, replace ρ_f with $\rho_W = 1.00 \times 10^3\,\text{kg/m}^3$ in the formula above to obtain the new value of the fraction. The answer is 8.0%.

9.89

The size of the raft is $A = 3.05\,\text{m} \times 6.10\,\text{m} = 18.605\,\text{m}^2$. As it settles further into the water by $\Delta h = 2.54\,\text{cm}$, the volume of the water it displaces increases by $\Delta V = A\Delta h$, which results in an increase in the buoyant force on the raft: $\Delta F_B = \rho_W g \Delta V = \rho_W g A \Delta h$. Thus the additional load ΔF_W the raft can take for each 2.54 cm it settles into the water is

$$\begin{aligned}\Delta F_W &= \Delta F_B = \rho_W g A \Delta h \\ &= (1000\,\text{kg/m}^3)(9.81\,\text{m/s}^2)(18.605\,\text{m}^2)(0.0254\,\text{m}) \\ &= 4.64 \times 10^3\,\text{N}\,.\end{aligned}$$

The weight of the raft (R) of size A and thickness h is $F_{WR} = \rho_R A h$, which must be balanced by the buoyant force F_B of the water. Suppose that the raft sinks into the water by a depth h', then $F_B = \rho_W A h'$. Equate F_W with F_B to obtain $\rho_R A h = \rho_W A h'$, which we solve for h':

$$h' = \frac{\rho_R h}{\rho_W} = \frac{(0.50\,\text{kg/m}^3)(0.305\,\text{m})}{1.000\,\text{kg/m}^3} = 0.15\,\text{m}\,.$$

9.95

The reduction in blood pressure at the location of the finger, ΔP, due to the change in the vertical position of the finger, Δh, is $\Delta P = \rho g \Delta h$, where ρ is the density of the blood. Numerically, this is

$$\Delta P = \rho g \Delta h = (1.05 \times 10^3\,\mathrm{kg/m^3})(9.81\,\mathrm{m/s^2})(-0.85\,\mathrm{m}) = -8.8\,\mathrm{kPa}\,,$$

which is comparable with the pressure P_H at the heart ($P_H = 13.3\,\mathrm{kPa}$). So the reduction in pressure is significant.

9.101

From Torricelli's Result [as well as Problem (9.97)] we know that the speed of the liquid jet emerging from the hole a distance h below the hole a distance h below the liquid surface is $v_2 = \sqrt{2gh}$. The direction of motion of the liquid jet is horizontal at point 2. The time-of-flight, t, of the jet can be found from $y = \frac{1}{2}gt^2$, or $t = \sqrt{2y/g}$. During that time the horizontal speed of the jet remains $v_2 = \sqrt{2gh}$. Thus the horizontal range of the liquid jet is

$$R = v_2 t = \sqrt{(2gh)\left(\frac{2y}{g}\right)} = 2\sqrt{hy}\,.$$

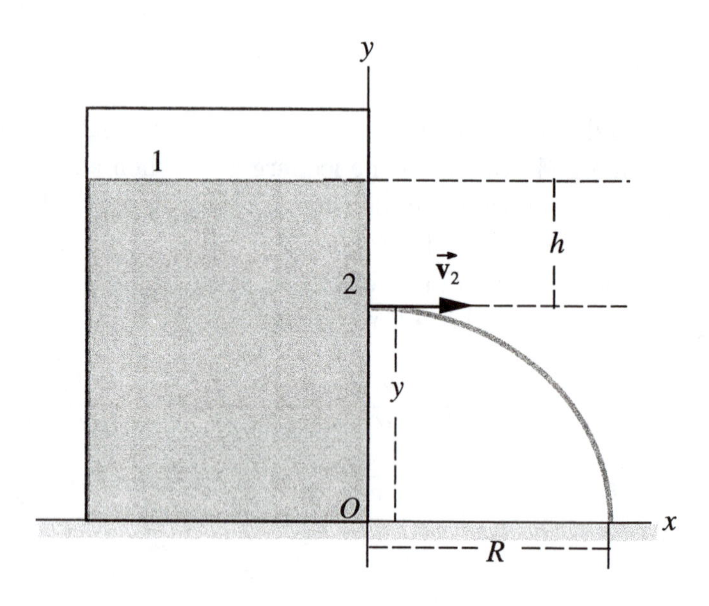

9.105

Examine the fluid flow depicted in Fig. P105. Since the fluid in both regions 2 and 3 originate at the same region (region 1), we may apply Bernoulli's Equation to regions 2 and 3:

$$P_2 + \frac{1}{2}\rho v_2^2 + \rho g y_2 = P_3 + \frac{1}{2}\rho v_3^2 + \rho g y_3\,.$$

Note that $P_2 = P$, $P_3 = P_0$, $v_2 = 0$, $v_3 = v$, and $y_2 = y_3$. So the equation above becomes $P = P_0 + \frac{1}{2}\rho v^2$. Solve for v:

$$v = \sqrt{\frac{2(P - P_0)}{\rho}} = \sqrt{\frac{2\Delta P}{\rho}}\,.$$

9.107

Apply Bernoulli's Equation to the flow at the pipe (p) and the throat (t): $P_p + \frac{1}{2}\rho v_p^2 + \rho g y_p = P_t + \frac{1}{2}\rho v_t^2 + \rho g y_t$. Note that $y_p = y_t$, and $A_p v_p = A_t v_t$ (the Continuity Equation), or $v_p = v_t(A_t/A_p)$. Thus the equation above reduces to

$$\Delta P = P_p - P_t = \frac{1}{2}\rho(v_t^2 - v_p^2) = \frac{1}{2}\rho v_t^2 \left(1 - \frac{A_t^2}{A_p^2}\right).$$

Now, the pressure difference ΔP between the pipe and the throat is responsible for the difference in the height of the two liquid columns, so $\Delta P = \rho g \Delta y$. Substitute this into the expression for ΔP above and solve for v_p:

$$v_p = \sqrt{\frac{2g\Delta y}{A_p^2/A_t^2 - 1}}.$$

9.109

When the vaccine fluid is inside the gun, its speed v_1 is zero while its pressure is $P_1 = (550\,\text{psi})(6.895 \times 10^4\,\text{Pa/psi}) = 3.792 \times 10^6\,\text{Pa}$, where we used the conversion factor $1.000\,\text{psi} = 6.895 \times 10^3\,\text{Pa}$ found in Problem (9.31). As the fluid is ejected into the air, its speed is v_2 and its pressure is $P_2 = P_A = 1.013 \times 10^5\,\text{Pa}$, since it is exposed to the air. For convenience, set $y_1 = y_2 = 0$ and apply Bernoulli's Equation:

$$P_1 + \frac{1}{2}\rho v_1^2 = P_1 = P_2 + \frac{1}{2}\rho v_2^2 = P_A + \frac{1}{2}\rho v_2^2,$$

which we solve for v_2:

$$v_2 = \sqrt{\frac{2(P_1 - P_A)}{\rho}} = \sqrt{\frac{2(3.792 \times 10^6\,\text{Pa} - 1.013 \times 10^5\,\text{Pa})}{1.1 \times 10^3\,\text{kg/m}^3}} = 82\,\text{m/s},$$

which amounts to about 180 mph.

9.113

Apply Bernoulli's Equation to points 1 and 2: $P_1 + \frac{1}{2}\rho v_1^2 + \rho g y_1 = P_2 + \frac{1}{2}\rho v_2^2 + \rho g y_2$. Note that, since the size of the tank is large, the speed v_1 of the water flow at the surface of the water in the tank is negligible [see Problem (9.102)]. Also, taking $y_2 = 0$ at the location of the hole, then $y_1 = H + h$. As for the pressures, we have $P_1 = P_3 = P_A$ and $P_2 = P_3 + \rho g h = P_A + \rho g h$. Substitute these expressions into Bernoulli's Equation to obtain

$$P_A + \rho g(H + h) = P_A + \rho g h + \frac{1}{2}\rho v_2^2.$$

Solve for v_2: $v_2 = \sqrt{2gH}$.

10 *Elasticity & Oscillation*

Answers to Selected Discussion Questions

•10.1

(a) low-carbon steel (b) tempered steel (c) all have the same slope, same Y (d) tempered steel (e) tempered steel (f) low-carbon steel (g) all are the same (h) low-carbon steel.

•10.3

(a) The area under the force-displacement curve is the work done on the sample, the energy mechanically entered into the material in the process of distorting it. (b) Since stress is force over area, and strain is displacement over length, the product of the two is the strain energy divided by the volume. It has the units of $\mathrm{J/m^3}$. (c) The colored area corresponds to the increased internal energy per unit volume of the sample.

•10.5

The stress in the broad base, which carries the entire load, is less, which is one reason why ancient walls tapered upward.

•10.7

Elephants have to walk around very carefully (they don't do much jumping). The compressive strength (Table 10.1) of a horse femur is only 145 MPa. We have bred horses to the point where they are almost too big for their bones — at least with all the jumping and running we demand of them.

•10.9

Human tendon has to be able to store a great deal of energy without permanently distorting — it's highly resilient, twice as much as spring steel. Bridges and springs must be resilient; they must recover after being compressed or stretched. Glass has very little ductility. The area under its stress-strain curve is small and it has little toughness — that's why a glass or china plate will shatter when dropped. A bow or pole should be resilient.

•10.11

When θ is a maximum $v = 0$, as is KE, a is a maximum as is PE. When $\theta = 0$, $a = 0$, as is PE, but v is maximum, as is KE.

•10.13

The parachute will swing like a pendulum and as the vortex frequency matches the pendulum frequency, the resulting resonance will send the rider wildly swinging.

•10.15

Like the bob on a helical spring, the greater the mass the smaller the frequency.

•10.17

At any moment (when at a distance r from the center of Mongo), the weight of the ball depends on the mass M of material remaining "beneath" it. Only the mass of the sphere of matter of radius r attracts the ball. $F_G \propto mM/r^2$, where M is proportional to the volume, which goes as r^3. Hence, $F_G \propto r$ and the motion must be SHM.

Answers to Odd-Numbered Multiple Choice Questions

1. d	**3.** c	**5.** b	**7.** b	**9.** a	**11.** c	**13.** b
15. b	**17.** a	**19.** b	**21.** c	**23.** d	**25.** c	

Solutions to Selected Problems

10.13

First, find the elastic constant k of the spring. Under a load of $F_1 = m_1 g$, where $m_1 = 2.00\,\text{kg}$, the spring stretches by $s_1 = 10.0\,\text{cm}$. Thus from $F_1 = ks_1$ we get

$$k = \frac{m_1 g}{s_1} = \frac{(2.00\,\text{kg})(9.81\,\text{m/s}^2)}{0.100\,\text{m}} = 196.2\,\text{N/m}\,.$$

When an additional mass m_2 ($= 0.50\,\text{kg}$) is attached to the first one the total force applied on the spring is $F = (m_1 + m_2)g$, so from $F = ks$ we find s, the amount of total elongation of the spring:

$$s = \frac{F}{k} = \frac{(m_1 + m_2)g}{k} = \frac{(2.00\,\text{kg} + 0.50\,\text{kg})(9.81\,\text{m/s}^2)}{196.2\,\text{N/m}} = 0.125\,\text{m}\,.$$

The additional amount of stretch for the spring is then $s - s_1 = 0.125\,\text{m} - 0.100\,\text{m} = 0.025\,\text{m}$.

10.23

The strain ε is defined by Eq. (10.5): $\varepsilon = \Delta L/L_0$. In this case $\Delta L = 1.5\,\text{cm} = 0.015\,\text{m}$ and $L_0 = 10\,\text{m}$, so

$$\varepsilon = \frac{\Delta L}{L_0} = \frac{0.015\,\text{m}}{10\,\text{m}} = 0.001\,5 = 0.15\%\,.$$

10.31

With twice the cross-sectional radius, the cross-sectional area A' of the new belt of radius R' is four times as much as A, the cross-sectional area of the old belt, whose radius is R: $A' = \pi R'^2 = \pi(2R)^2 = 4\pi R^2 = 4A$. Since the new belt is made of the same material, the stress σ_{max} that will cause it to break remains the same as before. So $\sigma_{\text{max}} = F/A = F'/A'$, where $F = 7363\,\text{N}$ and F' is the tensile load that will cause the new belt to break. Solve for F':

$$F' = F\left(\frac{A'}{A}\right) = (7363\,\text{N})\left(\frac{4A}{A}\right) = 29.5\,\text{kN}\,.$$

10.35

Since there are a total of five steel rivets, each of cross-sectional area $A = \frac{1}{4}\pi D^2$, with $D = 4.00\,\text{cm}$ the diameter of each rivet, the total cross-sectional area A_T under the shear stress is $A_T = 5A = 5\pi D^2/4$. To stay below the yield strength in shear of structural steel the shear force F that can be applied to the structure should cause a shear stress σ which does not exceed

σ_{max}: $\sigma = F/A_T \le \sigma_{max}$. Plug in $\sigma_{max} = 145\,\text{MPa}$ to find the maximum force F_{max} that can be applied:

$$F_{max} = \sigma_{max} A_T = \frac{5}{4}\pi D^2 \sigma_{max} = \frac{5}{4}\pi(0.0400\,\text{m})^2(145\times 10^6\,\text{Pa}) = 9.11\times 10^5\,\text{N}\,.$$

10.49

The force F which stretches the wire of length L_0 and cross-sectional area A is the weight of mass m, $F = F_W = mg$, which causes a stress $\sigma = F/A = mg/\pi R^2$ in the wire. Here R denotes the radius of the wire. The resulting strain in the wire is $\varepsilon = \Delta L/L_0$, where $\Delta L = 0.060\,\text{cm}$. Plug this value, together with $m = 4.00\,\text{kg}$, $R = 0.707\,\text{mm}$, and $L_0 = 60.0\,\text{cm}$, into Eq. (10.7) to find the Young's Modulus:

$$Y = \frac{\sigma}{\varepsilon} = \frac{mg/\pi R^2}{\Delta L/L_0} = \frac{mgL_0}{\pi R^2 \Delta L} = \frac{(4.00\,\text{kg})(9.81\,\text{m/s}^2)(6.00\,\text{cm})}{\pi(0.707\times 10^{-3})^2(0.060\,\text{cm})}$$

$$= 2.5\times 10^{10}\,\text{Pa} = 25\,\text{GPa}\,.$$

10.51

The maximum load F_{max} applied on the cable of cross-sectional area A is $F_{max} = 25\,000\,\text{lb}\times(4.454\,\text{N/lb}) = 1.113\times 10^5\,\text{N}$, which causes a stress of $\sigma = F_{max}/A = F/(\frac{1}{4}\pi D^2)$, with D the diameter of the cable. To stay below the yield strength σ_{max} of the cable, let $\sigma \le \sigma_{max} = F_{max}/A = F_{max}/(\frac{1}{4}\pi D^2_{min})$, which gives D_{min}, the required minimum diameter of the cable:

$$D_{min} = \sqrt{\frac{4F_{max}}{\pi\sigma_{max}}} = \sqrt{\frac{4(1.113\times 10^5\,\text{N})}{\pi(345\times 10^6\,\text{Pa})}} = 0.020\,3\,\text{m} = 2.03\,\text{cm}\,.$$

The corresponding maximum strain is then obtained from Eq. (10.7):

$$\varepsilon_{max} = \frac{\sigma_{max}}{Y} = \frac{345\,\text{MPa}}{200\,\text{GPa}} = 1.73\times 10^{-3} = 0.173\%\,.$$

10.61

The motion of the ant as seen by the child is a one-dimensional SHM, which is the circular motion of the ant projected onto a line perpendicular to the line-of-sight of the child. The maximum displacement of the ant either to the left or the right measured from the center of the motion, which coincides with the center of the record, is R, the radius of the record. Since the record turns at 78 rpm, the period of the motion is $T = (1/78)$ min, and the corresponding frequency is

$$f = \frac{1}{T} = \frac{78}{(1\text{min})(60\,\text{s/min})} = 1.3\,\text{Hz}\,.$$

According to Eq. (10.12) the angular frequency is $\omega = 2\pi f = (2\pi\,\text{rad})(1.3\,\text{s}^{-1}) = 8.2\,\text{rad/s}$.

10.79

The mechanical energy E of a spring-mass system consists of the kinetic energy KE of the mass m, given by KE $= \frac{1}{2}mv_x^2$, where $v_x = dx/dt$ is the speed of the mass; and the *elastic*-PE of the spring as it is stretched or compressed by an amount x: $\text{PE}_e = \frac{1}{2}kx^2$, with k the spring constant:

$$\text{E} = \text{KE} + \text{PE}_e = \frac{1}{2}mv_x^2 + \frac{1}{2}kx^2\,.$$

To observe the Conservation of Energy, set $d\text{E}/dt = 0$. The derivative for the KE term yields

$$\frac{d(\text{KE})}{dt} = \frac{d}{dt}\left(\frac{1}{2}mv_x^2\right) = \frac{1}{2}m\left(\frac{dv_x^2}{dt}\right) = \frac{1}{2}m\left(2v_x\frac{dv_x}{dt}\right) = mv_x\left(\frac{dv_x}{dt}\right) = mv_x\left(\frac{d^2x}{dt^2}\right)\,,$$

while that for the PE_e term gives

$$\frac{d(\text{PE}_e)}{dt} = \frac{d}{dt}\left(\frac{1}{2}kx^2\right) = \frac{1}{2}k\left(\frac{dx^2}{dt}\right) = \frac{1}{2}k\left(2x\frac{dx}{dt}\right) = kxv_x\,,$$

where we used $dx/dt = v_x$ and $dv_x/dt = d^2x/dt^2$. Thus

$$\frac{d\text{E}}{dt} = \frac{d(\text{KE})}{dt} + \frac{d(\text{PE}_e)}{dt} = mv_x\left(\frac{d^2x}{dt^2}\right) + kxv_x = 0\,.$$

Divide both sides of the last equality above by mv_x to yield

$$\frac{d^2x}{dt^2} + \frac{k}{m}x = 0\,.$$

10.85

The weight of the potatoes of mass m is $F_W = mg$, which is the force F applied on the scale. The resulting displacement of the scale is $x = 2.50\,\text{cm} = 0.025\,0\,\text{m}$; hence from $F_W = mg = F = kx$ we get

$$k = \frac{mg}{x} = \frac{(2.00\,\text{kg})(9.81\,\text{m/s}^2)}{0.025\,0\,\text{m}} = 785\,\text{N/m}\,.$$

The frequency f_0 of the SHM is found from Eq. (10.20):

$$f_0 = \frac{1}{2\pi}\sqrt{\frac{k}{m}} = \frac{1}{2\pi}\sqrt{\frac{785\,\text{N/m}}{2.00\,\text{kg}}} = 3.15\,\text{Hz}\,.$$

10.91

Let the mass of the object be m, then its weight is $F_w = mg$ which, when applied to the spring, results in an elongation of $x = 2.0\,\mathrm{cm}$. Thus the spring constant k is given by $k = mg/x$. The frequency f_0 of the subsequent SHM then follows from Eq. (10.20):

$$f_0 = \frac{1}{2\pi}\sqrt{\frac{k}{m}} = \frac{1}{2\pi}\sqrt{\frac{mg/x}{m}} = \frac{1}{2\pi}\sqrt{\frac{g}{x}} = \frac{1}{2\pi}\sqrt{\frac{9.81\,\mathrm{m/s^2}}{0.020\,\mathrm{m}}} = 3.5\,\mathrm{Hz}\,.$$

11 *Waves & Sound*

Answers to Selected Discussion Questions

•11.1

The waves vanish, which certainly means there is no *elastic*-PE stored in the undistorted rope. All the energy is kinetic, though undisplaced at that very instant the various segments of the rope in the region of overlap are nonetheless moving vertically.

•11.3

The speed of the wavepulse varies with the square root of the tension, which, in turn, is determined by the load and the weight of the string itself. The tension increases all the way up to the point of support because of the linear mass-density of the string. Hence, the speed increases as the wave rises. Air friction and internal losses will convert some of the energy of the wave to thermal energy and its amplitude will diminish.

•11.5

The materials are getting progressively more rigid as we go from clay down the table. Just compare the softness of clay to the hardness of granite. Accordingly, we can expect that the internal restoring force will increase as the interatomic force increases, and so the speed of a compression wave will increase. Of course, an increase in rigidity corresponds to an increase in Young's Modulus.

•11.7

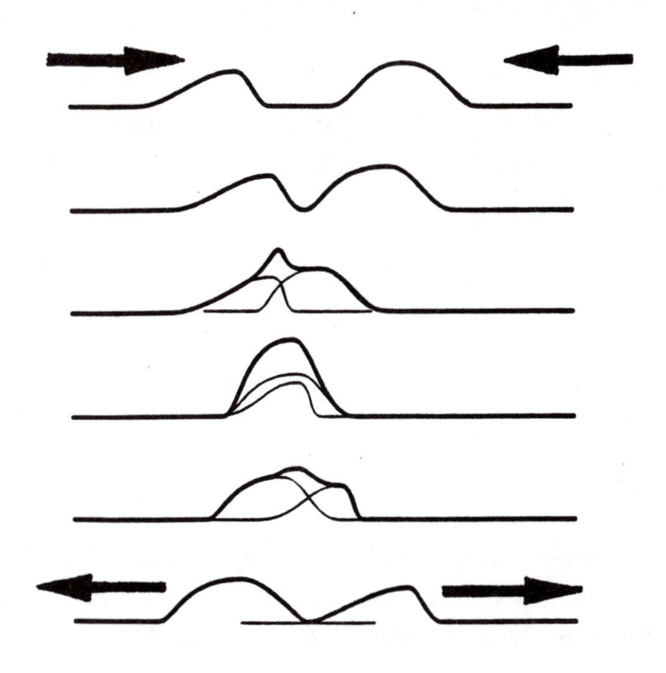

•11.9

The strings warm from the friction — they expand. There is a decrease in tension and a drop in speed (v) and frequency (f). The wind instruments warm from the breath, increasing v and therefore f.

•11.13

The answer lies in Young's Law of Strings (1800), which was discussed though not under its formal name. Remember that the point at which the string is struck must be an antinode because it is vibrating with maximum amplitude there. But the 7th harmonic has 7 antinodes and divides the string into 7 equal-length segments. Thus, 1/7th of the way down the string oscillating in the 7th harmonic, there must be a node. Hence, if we strike a string at that 1/7th distance, we preclude the presence of the unpleasant 7th harmonic.

•11.21

The triangular wave contains many strong overtones or Fourier components and so has a much greater high-frequency content, and sounds more "metallic". The sharper the bends in the string, the more high-frequency terms will be present. Plucking with the fingers will therefore generate fewer overtones and sound more mellow.

Answers to Odd-Numbered Multiple Choice Questions

1. c	**3.** d	**5.** b	**7.** c	**9.** e	**11.** d	**13.** d
15. b	**17.** c	**19.** a	**21.** b	**23.** c		

Solutions to Selected Problems

11.5

For a periodic wave of wavelength λ, wave speed v and frequency f, Eq. (11.1) holds true: $v = f\lambda$. Rewrite this as $f = v/\lambda$ and multiply both sides by 2π to obtain $2\pi f = 2\pi v/\lambda = (2\pi/\lambda)v$. But $2\pi f = \omega$ (the angular frequency of the wave), so $\omega = (2\pi/\lambda)v$.

11.13

We are given a formula for $\partial z/\partial x$, and are looking for an expression for $\partial x/\partial t$. So let $z \to x$ and $x \to t$ to rewrite the given formula for $\partial z/\partial x$ as

$$\left(\frac{\partial x}{\partial t}\right)_F = -\frac{\left(\frac{\partial F}{\partial t}\right)_x}{\left(\frac{\partial F}{\partial x}\right)_t}.$$

Comparing with the expression for v_x, which requires that φ be kept constant while taking $\partial x/\partial t$, it is clear that F here should be φ $(= kx - \omega t)$. In Problems (11.11) and (11.12) we already found

$$\left(\frac{\partial \varphi}{\partial t}\right)_x = \frac{\partial}{\partial t}(kx - \omega t) = -\omega$$

and

$$\left(\frac{\partial \varphi}{\partial x}\right)_t = \frac{\partial}{\partial x}(kx - \omega t) = k,$$

which we substitute into the expression for $\partial x/\partial t$ obtained above, with F replaced by φ, to find the phase speed v_x:

$$v_x = \left(\frac{\partial x}{\partial t}\right)_\varphi = -\frac{\left(\frac{\partial \varphi}{\partial t}\right)_x}{\left(\frac{\partial \varphi}{\partial x}\right)_t} = -\frac{-\omega}{k} = \frac{\omega}{k}.$$

11.19

Think of the wave as having been formed by raising a rectangular portion of the water (of mass m and height A) out, to form the trough; and up, to form the crest. This amounts to lifting a mass m through a vertical displacement A, requiring an amount of energy mgA. When the amplitude of the wave is doubled, twice as much water has to be moved up: $m \to 2m$; and the vertical displacement is now $2A$ instead of A. Thus the amount of energy needed to form the new wave with twice the amplitude is

$$(2m)g(2A) = 4mgA\,,$$

which is 4 times as much as before.

11.31

Suppose that the picture in the text (Fig. P31) was taken at $t = 0$. In the following sequence, the top figure depicts the situation a little before $t = 2\,\mathrm{s}$, the middle one is the picture at exactly $t = 2\,\mathrm{s}$ (when the two peaks coincide), while the bottom one is at $t = 4\,\mathrm{s}$.

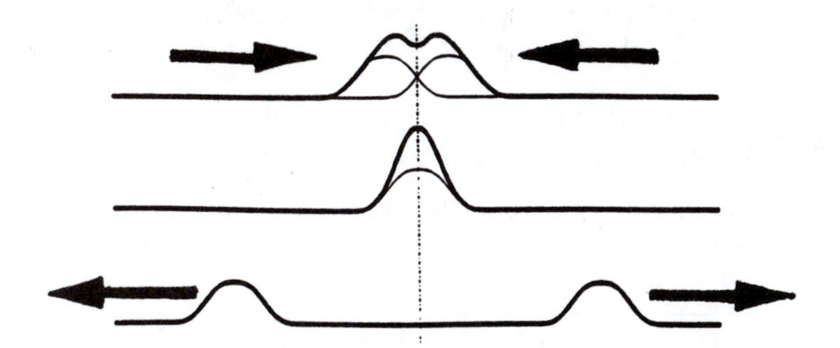

11.35

If the amplitude of a harmonic wave is A and its frequency is f, then each particle in the medium carrying the wave is in a simple harmonic motion with the same frequency and amplitude. According to Eq. (10.17) the maximum acceleration of each particle in the wave is then $a_{\max} = [-\omega^2 A\cos\omega t]_{\max} = \omega^2 A = (2\pi f)^2 A$. In this case, the waveform is $y = 0.040\sin(2\pi x)$ which, when compared with the standard form in Eq. (11.2), $y = A\sin(2\pi x/\lambda)$, yields $A = 0.040\,\mathrm{m}$ and $\lambda = 1.0\,\mathrm{m}$. Thus

$$a_{\max} = (2\pi f)^2 A = \left(\frac{2\pi v}{\lambda}\right)^2 A = \left[\frac{2\pi(2.0\,\mathrm{m/s})}{1.0\,\mathrm{m}}\right]^2 (0.040\,\mathrm{m}) = 6.3\,\mathrm{m/s^2}\,,$$

where we used $f = v/\lambda$.

11.39

The wave speed v is obtained from Eq. (11.4): $v = \sqrt{F_T/\mu}$. Here $F_T = 80\,\mathrm{N}$ is the tension in the string and $\mu = 7.5\,\mathrm{g/m} = 7.5\times 10^{-3}\,\mathrm{kg/m}$ is the linear mass-density:

$$v = \sqrt{\frac{F_T}{\mu}} = \sqrt{\frac{80\,\mathrm{N}}{7.5\times 10^{-3}\,\mathrm{kg/m}}} = 1.0\times 10^2\,\mathrm{m/s}\,.$$

Since $v \propto \sqrt{F_T}$, $F_T \propto v^2$. Thus if v is to be doubled the tension should be increased to $F'_T = (2v/v)^2 F_T = 4F_T = 4(80\,\text{N}) = 3.2 \times 10^2\,\text{N} = 0.32\,\text{kN}$.

11.43

Since the rope is uniform the mass Δm of a segment of length Δy is proportional to Δy: $\Delta m/m = \Delta y/L$, so $\Delta m = m\Delta y/L$.

Now consider the portion of the rope below the height Δy. The tension F_T at height Δy should support the weight F_W of that portion of the rope:

$$F_T = F_W = (\Delta m)g = \frac{mg\Delta y}{L}.$$

From Eq. (11.4), the speed v of the wave at height Δy is therefore

$$v = \sqrt{\frac{F_T}{\mu}} = \sqrt{\frac{mg\Delta y/L}{\mu}} = \sqrt{g\Delta y},$$

which increases with Δy. The maximum speed occurs at $\Delta y = L$ (the top of the string), where $v = v_{max} = \sqrt{gL}$. (This is understandable since F_T is the greatest over there.)

If a mass M is hung at the bottom of the rope then the tension F_T at any height in the rope will have to increase by Mg to support the additional weight of that mass: $F_T = mg\Delta y/L + Mg$. This gives

$$v = \sqrt{\frac{F_T}{\mu}} = \sqrt{\frac{mg\Delta y/L + Mg}{\mu}} = \sqrt{g\Delta y + \frac{MgL}{m}}.$$

11.59

The wavelength λ of the wave on the string is given by

$$\lambda = \frac{v}{f} = \frac{344\,\text{m/s}}{1000\,\text{Hz}} = 0.344\,\text{m}.$$

Thus the number of wavelengths in 1 m is $1\,\text{m}/0.344\,\text{m} = 2.91$.

11.79

Use Eq. (11.8), $I = \text{P}/A$. Here $\text{P} = 50\,\text{W}$, and $A = 4\pi R^2$, with $R = 10\,\text{m}$; so

$$I = \frac{\text{P}}{4\pi R^2} = \frac{50\,\text{W}}{4\pi(10\,\text{m})^2} = 0.040\,\text{W/m}^2 = 40\,\text{mW/m}^2.$$

Since I is the power passing through a unit cross-sectional area, the power intercepted by a detector of area A' is $\text{P}' = IA'$, and so the energy E that passes through the detector during a time interval Δt is $\mathrm{E} = \mathrm{P}'\Delta t = IA'\Delta t$. Plug in $A = (1.0\,\mathrm{cm}^2)(10^{-2}\,\mathrm{m/cm})^2 = 1.0 \times 10^{-4}\,\mathrm{m}^2$ and $\Delta t = 1.0\,\mathrm{s}$ to obtain

$$\mathrm{E} = \mathrm{P}'\Delta t = IA'\Delta t = (0.040\,\mathrm{W/m}^2)(1.0 \times 10^{-4}\,\mathrm{m}^2)(1.0\,\mathrm{s}) = 4.0 \times 10^{-6}\,\mathrm{J} = 4.0\,\mu\mathrm{J}\,.$$

11.95

The sound-level β is related to the intensity I by Eq. (11.10): $\beta = 10\log_{10}(I/I_0)$. Since the two systems have the same intensity at the location of the microphone, they must also have the same sound-level. The difference in their sound-levels is therefore 0 dB.

11.99

Rewrite Eq. (11.10), $\beta = 10\log_{10}(I/I_0)$, as $\beta/10 = \log_{10}(I/I_0)$. Now raise 10 to the power of both sides to obtain

$$10^{\beta/10} = 10^{\log_{10}(I/I_0)} = \frac{I}{I_0}\,,$$

where in the last step we made use of the algebraic identity $10^{\log_{10} x} = x$. It follows that $I = 10^{\beta/10} I_0$. In this case $\beta = 77\,\mathrm{dB}$, so

$$I = 10^{\beta/10} I_0 = \left(10^{77/10}\right)(10^{-12}\,\mathrm{W/m}^2) = 5.0 \times 10^{-5}\,\mathrm{W/m}^2\,.$$

11.109

The sound intensity I varies as the square of the acoustic pressure P: $I \propto P^2$. Thus $I/I_0 = (P/P_0)^2$, where $P_0 = 2 \times 10^{-5}\,\mathrm{Pa}$ corresponds to an intensity of I_0. Thus from Eq. (11.10)

$$\beta = 10\log_{10}\left(\frac{I}{I_0}\right) = 10\log_{10}\left(\frac{P}{P_0}\right)^2 = 20\log_{10}\left(\frac{P}{P_0}\right).$$

11.115

According to the problem statement the period of the beats is $T_{\text{beat}} = 0.99\,\mathrm{s}$, so the beat frequency is $f_{\text{beat}} = 1/T_{\text{beat}} = 1/0.99\,\mathrm{s} = 1.0\,\mathrm{Hz}$, which is the same as Δf, the difference in frequency between the two tuning forks which produce the beats.

11.119

First, find the speed v of the wave on the piano string of length L and mass m under tension F_T using Eq. (11.4), $v = \sqrt{F_T/\mu}$, where μ is the linear mass-density of the string:

$$v = \sqrt{\frac{F_T}{\mu}} = \sqrt{\frac{100\,\mathrm{N}}{2.5 \times 10^{-3}\,\mathrm{kg}/1.00\,\mathrm{m}}} = 2.0 \times 10^2\,\mathrm{m/s}\,.$$

Also, the wavelength λ_1 of the fundamental mode is given by Eq. (11.14), with $N = 1$: $\lambda_1 = 2L/1 = 2(1.00\,\text{m})/1 = 2.00\,\text{m}$. Thus the corresponding frequency f_1 is

$$f_1 = \frac{v}{\lambda_1} = \frac{2.0 \times 10^2\,\text{m/s}}{2.00\,\text{m}} = 1.0 \times 10^2\,\text{Hz} = 0.10\,\text{kHz}\,.$$

11.123

For $N = 1$ the standing-wave displacement is given by

$$y_1(x,t) = A_1 \sin\frac{\pi x}{L}\cos 2\pi f_1 t\,.$$

Recall that a node is where y remains zero for all t. Thus the nodes of y_1 must satisfy $\sin(\pi x/L) = 0$, or $\pi x/L = n\pi$, where n is an integer. Thus the nodes for y_1 are located at $x = nL$. Since the string runs from $x = 0$ to $x = L$ the only possibilities are $n = 0$ and $n = 1$, corresponding to $x = 0$ and $x = L$, respectively. In fact these are the two ends of the string. There are no nodes in between. The antinodes are where the amplitude of y_1 is at its maximum, with $|\sin(\pi x/L)| = 1$. For $0 \le x \le L$ This gives $\pi x/L = \pi/2$, or $x = L/2$, the middle of the string. Note that the antinode is midway in between the two nodes. See the diagram below.

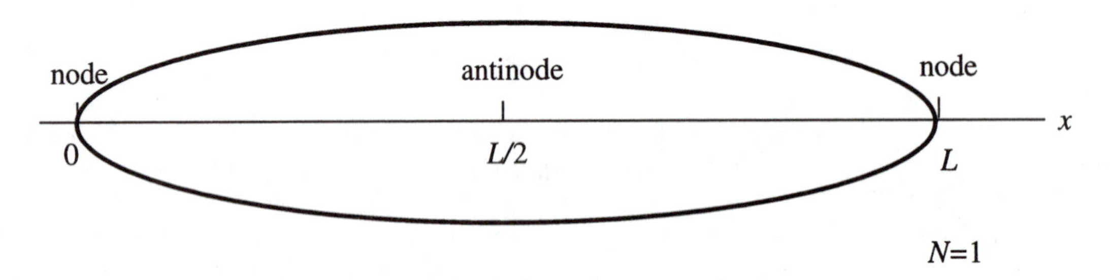

Similarly, the nodes for the $N = 2$ mode must satisfy $\sin(2\pi x/L) = 0$, or $2\pi x/L = n\pi$, whereupon $x = nL/2$. For $0 \le x \le L$ the possible integral values of n are 0, 1 and 2, and so the nodes for y_2 are at $x = 0$, $L/2$ and L. For the antinodes we require that $|\sin(2\pi x/L)| = 1$, or $2\pi x/L = \pi/2$ or $3\pi/2$ for the range $0 \le x \le L$. Thus $x = L/4$ or $3L/4$. Once again, each antinode are always located midway between two nodes. See the diagram below.

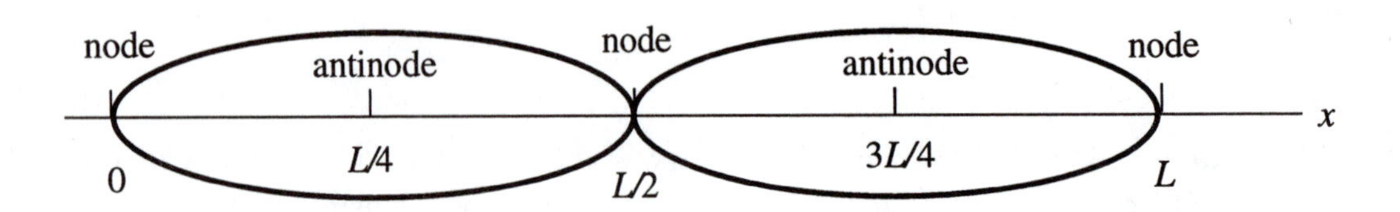

In general, for the N-th standing wave mode on a string of length L with both ends fixed there must be an integral (N) number of "loops", each running from one node to the next with a length of $\lambda/2$:

$$L = N\frac{\lambda}{2}\,.$$

Thus $N = 2L/\lambda$, which we plug into the expression for $y(x,t)$:

$$\begin{aligned} y(x,t) &= A_{\rm N}\sin\frac{N\pi x}{L}\cos 2\pi f_{\rm N}t \\ &= A_{\rm N}\sin\frac{(2L/\lambda)\pi x}{L}\cos 2\pi f_{\rm N}t \\ &= A_{\rm N}\sin kx\cos 2\pi f_{\rm N}t\,, \end{aligned}$$

where in the last step we used the definition $k = 2\pi/\lambda$.

The transverse velocity of a point on the string located at x is the time-rate-of-change of y, the transverse displacement:

$$\begin{aligned} v_y = \left(\frac{\partial y}{\partial t}\right)_x &= \frac{\partial}{\partial t}\left(A_{\rm N}\sin kx\cos 2\pi f_{\rm N}t\right) \\ &= A_{\rm N}\sin kx\,\frac{d}{dt}\left(\cos 2\pi f_{\rm N}t\right) \\ &= -2\pi f_{\rm N}A_{\rm N}\sin kx\,\sin 2\pi f_{\rm N}t\,. \end{aligned}$$

The corresponding speed is $v_y = |v_y| = 2\pi f_{\rm N}A_{\rm N}\,|\sin kx\,\sin 2\pi f_{\rm N}t|$.

11.129

The wavelength λ of the standing sound wave in a narrow pipe is solely determined by the length of the pipe, which does not change as it is filled with helium in stead of air. Thus from $f = v/\lambda$ we know that $f \propto v$, the speed of sound in the pipe. This gives $f_{\rm He}/f_{\rm air} = v_{\rm He}/v_{\rm air}$. According to Table (11.2) the speed of sound in helium at 0℃ is $v_{\rm He} = 970\,{\rm m/s}$, while that in the air is $331.45\,{\rm m/s}$; so when it is filled with helium the frequency of the organ will be

$$f_{\rm He} = f_{\rm air}\left(\frac{v_{\rm He}}{v_{\rm air}}\right) = (600\,{\rm Hz})\left(\frac{970\,{\rm m/s}}{331.45\,{\rm m/s}}\right) = 1.76\,{\rm kHz}\,.$$

11.145

The beat frequency, $f_{\rm beat}$, is the difference between $f_{\rm s}$, the source frequency; and $f_{\rm o}$, the observed frequency of the reflected signal. According to Eq. (11.23) $f_{\rm o} = f_{\rm s}(v + v_{\rm t})/(v - v_{\rm t})$, where $v_{\rm t}$ is the speed of the target approaching the observer. So the beat frequency is

$$f_{\rm beat} = f_{\rm o} - f_{\rm s} = f_{\rm s}\left(\frac{v + v_{\rm t}}{v - v_{\rm t}}\right) - f_{\rm s} = \frac{2v_{\rm t}f_{\rm s}}{v - v_{\rm t}}\,.$$

12 *Thermal Properties of Matter*

Answers to Selected Discussion Questions

•12.1

Here are a few: (a) The bore hole in the glass tube must be uniform. (b) All of the mercury, including the stuff way up in the stem, must be at the same temperature. (c) Generally, to speed up its response the walls of the bulb are made thin and that thinness makes it vulnerable to pressure variations, which change its volume (via barometric changes or hydrostatic pressure, if it's immersed in a liquid). (d) There is a variation in pressure in the mercury due to the different heights of the column. (e) There is a difference in internal pressure if it's held vertically as opposed to horizontally. (f) There are errors associated with the softness of the glass. If the thermometer is raised to a high temperature and then cooled rapidly, it might take weeks for the glass to return to its original volume. Try measuring the freezing point of water before and immediately after reading its boiling point — the difference can be as great as $1\,\mathrm{C}^\circ$. (g) The mere presence of the thermometer in a small system may change the temperature of the system. (h) When measuring a changing temperature at any moment, the thermometer will always read warmer if the bath temperature is falling and vice versa.

•12.3

We know that the antimony expands on solidifying, as does water. Since that would be a very helpful trait for a casting material to have (it would fill all the fine details in the mold), it's reasonable to expect that's the answer to this question.

•12.7

Being a poor conductor, the center of a boulder so treated would remain quite hot while the outside was cooled and contracted rapidly. Pressure would build up, there would be considerable internal stress, and the thing would rupture at any flaw or weak spot.

•12.9

Bulbs are cheaper thinner and can tolerate the changes in temperature associated with ordinary operation, which are fairly gradual. They are made of inexpensive glass with a relatively large β, so a drop of water (or latex paint) can cause enough contraction and stress to shatter a hot light bulb. Clearly, outdoor lamps have to be protected from rain and snow. The heating in a flashbulb is so rapid even the thin walls tend to burst and they therefore are usually enclosed in a tough plastic film to keep them from shattering.

•12.11

At a temperature of even $-1\,^{\circ}\mathrm{C}$ it takes about 140 atm of pressure ($\approx 2000\,\mathrm{lb/in^2}$) to melt ice, so the problem was probably that the snow was just too cold.

•12.13

The liquid will expand rapidly: its density decreasing as the density of the vapor increases. The surface meniscus will flatten out and then disappear altogether when the density of liquid and vapor are equal at a pressure of 7.38 MPa.

•12.15

Yes. They were called permanent because they could not at first be liquified, which suggests a weak intermolecular cohesive force and that, in turn, suggests ideal behavior.

•12.17

$PV = nRT$, so both pressures must be equal since everything else is the same. The speeds of the hydrogen molecules must be greater than those of the nitrogen because the average KE is the same from Eq. (12.15). The pressures can be equal because the lighter hydrogens hit the walls at greater speeds and they do it more frequently because they traverse the chamber more quickly. The pressure is proportional to the average KE via Eq. (12.14).

•12.19

Remember that the pressure in the room is more or less constant. Increasing T increases the average KE, which increases the net KE and the P but leads to an over-pressure and an outward current of air. The room leaks warm air to the outside. The temperature goes up because it's dependent on the average KE of each molecule, which is higher. The pressure remains the same because it depends on both the number of molecules per unit volume and their average KE. Compare Eqs. (12.14) and (12.15).

Answers to Odd-Numbered Multiple Choice Questions

1. b **3.** a **5.** c **7.** a **9.** b **11.** a **13.** a
15. b **17.** e **19.** a

Solutions to Selected Problems

12.9

First convert the temperatures to Celsius scale. For $T_F = 0\,°\mathrm{F}$, $T_C = \frac{5}{9}(T_F - 32\,°\mathrm{C}) = \frac{5}{9}(0 - 32)\,°\mathrm{C} = -17.8\,°\mathrm{C}$; and for $T_F = 100\,°\mathrm{F}$, $T_C = \frac{5}{9}(T_F - 32) = \frac{5}{9}(100 - 32)\,°\mathrm{C} = 37.8\,°\mathrm{C}$. Thus the temperature change, in Celsius, is $\Delta T_C = 37.8\,°\mathrm{C} - (-17.8\,°\mathrm{C}) = 55.6\,\mathrm{C}°$. The corresponding temperature change in Kelvins is the same as that in Celsius: $\Delta T = 55.6\,\mathrm{K}$.

12.19

Apply Eq. (12.3), $\Delta L = \alpha L_0\,\Delta T$, which gives the change in length of an object of length L_0 as a result of ΔT, a change in its temperature. Here α is the coefficient of linear expansion of the material of which the object is made. In this case we are dealing with an aluminum bar, for which $\alpha = 25 \times 10^{-6}\,\mathrm{K}^{-1}$, $\Delta T = 50\,°\mathrm{C} - 30\,°\mathrm{C} = 20\,\mathrm{C}° = 20\,\mathrm{K}$, and $L_0 = 10\,\mathrm{m}$. So the length of the bar will increase by

$$\Delta L = \alpha L_0\,\Delta T = (25 \times 10^{-6}\,\mathrm{K}^{-1})(10\,\mathrm{m})(20\,\mathrm{K}) = 5.0 \times 10^{-3}\,\mathrm{m} = 5.0\,\mathrm{mm}\,.$$

12.29

Apply Eq. (12.4) for volume expansion: $\Delta V = \beta V_0\,\Delta T$. Here $\beta = 182 \times 10^{-6}\,\mathrm{K}^{-1}$ is the coefficient of volume expansion for mercury, $V_0 = 0.50\,\mathrm{cm}^2$ is its initial volume at $32\,°\mathrm{F}$, and $\Delta T = \Delta T_C = \frac{5}{9}\,\Delta T_F = \frac{5}{9}(T_{Ff} - T_{Fi}) = \frac{5}{9}(212 - 32) = 100\,\mathrm{K}$. (Note that $1\,\mathrm{K} = \frac{9}{5}\,\mathrm{F}°$ for temperature change measured in K and in F°.) Thus the new volume at $212\,°\mathrm{F}$ is given by

$$\begin{aligned} V &= V_0 + \Delta V = V_0(1 + \beta\,\Delta T) \\ &= (0.50\,\mathrm{cm}^3)\left[1 + (182 \times 10^{-6}\,\mathrm{K}^{-1})(100\,\mathrm{K})\right] \\ &= 0.51\,\mathrm{cm}^3\,. \end{aligned}$$

12.39

The period τ_0 of the pendulum of length L_0 at $20.00\,°\mathrm{C}$ is given by

$$\tau_0 = 2\pi\sqrt{\frac{L_0}{g}} = 2\pi\sqrt{\frac{1.000\,\mathrm{m}}{9.81\,\mathrm{m/s^2}}} = 2.006\,\mathrm{s}\,.$$

When the temperature drops by $\Delta T = 0.000\,^\circ\mathrm{C} - 20.00\,^\circ\mathrm{C} = -20.00\,\mathrm{C}^\circ = -20.00\,\mathrm{K}$, the length of the pendulum changes to $L = L_0 + \Delta L = L_0 + \alpha L_0\,\Delta T$ (where α is the coefficient of linear expansion for aluminum); and so the new period is

$$\begin{aligned}\tau &= 2\pi\sqrt{\frac{L}{g}} = 2\pi\sqrt{\frac{L_0(1+\alpha\,\Delta T)}{g}}\\ &= 2\pi\sqrt{\frac{1.000\,\mathrm{m}\,[1+(25\times10^{-6}\,\mathrm{K}^{-1})(-20\,\mathrm{K})]}{9.81\,\mathrm{m/s^2}}}\\ &= 2.006\,\mathrm{s} < \tau_0\,,\end{aligned}$$

meaning that the clock runs slightly faster.

12.51

Apply Eq. (12.8) to the initial state of the helium gas (labeled with subscript i) and the final state at STP (labeled with f): $P_i V_i/T_i = P_f V_f/T_f = \text{constant}$. Here $P_i = 99\,\mathrm{kPa}$, $V_i = 1200\,\mathrm{cm^3}$, $T_i = 273.15 + 15 = 288.15\,\mathrm{K}$, $P_f = 1.00\,\mathrm{atm} = 101\,\mathrm{kPa}$, and $T_f = 273.15\,\mathrm{K}$. Solve for the final volume of the gas:

$$V_f = V_i\left(\frac{P_i}{P_f}\right)\left(\frac{T_f}{T_i}\right) = (1200\,\mathrm{cm^3})\left(\frac{99\,\mathrm{kPa}}{101\,\mathrm{kPa}}\right)\left(\frac{(273.15\,\mathrm{K}}{288.15\,\mathrm{K}}\right) = 1.1\times10^3\,\mathrm{cm^3}\,.$$

12.55

Each oxygen molecule, O_2, is made of two oxygen atoms, each of mass 15.999 u [see Problem (12.53)]. Thus its molecular mass is

$$m = [2(15.999\,\mathrm{u/molecule})]\,(1.6606\times10^{-27}\,\mathrm{kg/u}) = 5.314\times10^{-26}\,\mathrm{kg/molecule}\,,$$

and the number of oxygen molecules with a total mass of $M = 16.0\,\mathrm{kg}$ is then

$$N = \frac{M}{m} = \frac{16.0\,\mathrm{kg}}{5.314\times10^{-26}\,\mathrm{kg/molecule}} = 3.01\times10^{26}\,\mathrm{molecules}\,,$$

which corresponds to

$$n = \frac{N}{N_A} = \frac{3.011\times10^{26}\,\mathrm{molecules}}{6.022\times10^{23}\,\mathrm{molecules/mol}} = 500\,\mathrm{mol}\,.$$

14.61

(a) Refer to the free-body diagram for the piston shown to the right. The piston is subject to two forces: its weight, mg, downward; and the force due to the pressure of the gas, PA, upward. For the piston to achieve mechanical equilibrium $+\uparrow\sum F = PA - mg = 0$, so

$$P = \frac{mg}{A}.$$

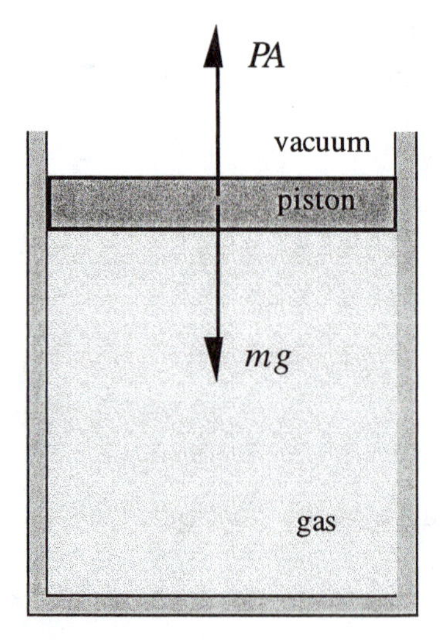

(b) As the piston moves upward by an amount Δy the volume of the gas in the cylinder increases by $\Delta V = A\,\Delta y$. Since the temperature of the gas is fixed $PV = C$, and so P must decrease as V increases by ΔV. This causes a reduction in the upward force, PA, resulting in a net downward force ΔF on the piston. To find the change in pressure, ΔP, as a result of a small ΔV we must find the relationship between dP and dV. To do so, start from Boyle's Law, $PV = C$, and differentiate both sides: $d(PV) = PdV + VdP = dC = 0$, or $dP = -(P/V)dV$. Thus for a finite yet small change in V the corresponding change in P is approximately

$$\Delta P \approx -\frac{P}{V}\,\Delta V = -\frac{mg/A}{V}(A\,\Delta y) = -\frac{mg\,\Delta y}{V},$$

and, choosing up as positive, the net force on the piston is now

$$\Delta F = (P + \Delta P)A - mg = A\,\Delta P = A\left(-\frac{mg\,\Delta y}{V}\right) = -\frac{mgA\,\Delta y}{V},$$

where the negative sign indicated that the net force on the piston is downward as it is raised from its equilibrium position.

12.71

Label the oxygen gas with subscript O and the nitrogen with N, respectively, and consider the two gases separately. The expansion process for either gas is isothermal (i.e., T = constant.), so for the oxygen gas Eq. (12.5) gives $P_{\text{iO}}V_{\text{iO}} = P_{\text{fO}}V_{\text{fO}}$, which yields P_{fO}, the final pressure in the chamber due to the oxygen gas alone:

$$P_{\text{fO}} = \frac{P_{\text{iO}}V_{\text{iO}}}{V_{\text{fO}}} = \frac{(5.0\,\text{atm})(3.0\,\text{liters})}{4.0\,\text{liters}} = 3.75\,\text{atm}.$$

Similarly for the nitrogen gas $P_{\text{iN}}V_{\text{iN}} = P_{\text{fN}}V_{\text{fN}}$, so P_{fN}, the contribution from the nitrogen to the final pressure in the chamber, is

$$P_{\text{fN}} = \frac{P_{\text{iN}}V_{\text{iN}}}{V_{\text{fN}}} = \frac{(2.0\,\text{atm})(1.0\,\text{liters})}{4.0\,\text{liters}} = 0.50\,\text{atm}.$$

The total pressure P_f in the chamber is the sum of the contributions from both gases: $P_f = P_{fO} + P_{fN} = 3.75\,\text{atm} + 0.50\,\text{atm} = 4.3\,\text{atm}$.

12.91

Use Eq. (12.16), $v_{rms} = \sqrt{3k_B T/m}$, to find v_{rms}. For oxygen molecules with a molecular mass of $m = 5.3136 \times 10^{-26}\,\text{kg}$ at $T = 293.15\,\text{K}$,

$$v_{rms} = \sqrt{\frac{k_B T}{m}} = \sqrt{\frac{3(1.380662 \times 10^{-23}\,\text{J/K})(293.15\,\text{K})}{5.3136 \times 10^{-26}\,\text{kg}}} = 478.03\,\text{m/s}.$$

12.99

Start with Eq. (12.14), $PV = \frac{2}{3}N(\text{KE})_{av}$, by rewriting $(\text{KE})_{av}$ on the RHS as $(\text{KE})_{av} = \frac{1}{2}mv^2_{rms}$, where m is the molecular mass:

$$PV = \frac{2}{3}N(\text{KE})_{av} = \frac{2}{3}N\left(\frac{1}{2}mv^2_{rms}\right) = \frac{1}{3}(Nm)v^2_{rms}.$$

Since Nm is the total mass of the N molecules, the density of the gas of volume V is given by $\rho = mN/V$. Use this result to solve for v_{rms} from the equation above:

$$v_{rms} = \sqrt{\frac{3PV}{mN}} = \sqrt{\frac{3P}{mN/V}} = \sqrt{\frac{3P}{\rho}}.$$

13 *Heat & Thermal Energy*

Answers to Selected Discussion Questions

•13.3

Bicycling consumes the same amount of energy per hour as shivering but converts 80% of it into thermal energy. Shivering, which doesn't perform any work to speak of, converts chemical energy almost completely into thermal energy. We shiver when the body loses too much heat and needs a quick infusion of thermal energy.

•13.5

Copper is a much better conductor than iron or steel, but iron is cheaper and chemically a better material to cook food on. The copper layer is there to rapidly and uniformly distribute the heat by conduction. To the same end, good pots are generally made thick-bottomed, but that's not necessary if one just wants to boil water.

•13.7

Q/t should be proportional to the exposed area A, the temperature difference between the body and the fluid ΔT, and the physical parameters that describe the particular situation, such as geometry, orientation, wind speed, etc. All of the latter are combined into a constant of proportionality, $k_{\rm C}$, called the convection coefficient, which is determined experimentally. Hence, $Q/t = k_{\rm C} A \Delta T$. For a human on a windless day, $k_{\rm C} \approx 5\,\text{kcal/m}^2\cdot\text{h}\cdot\text{C}^\circ$, and A is the exposed area of the skin.

•13.9

We tend to be more comfortable in a dry-air environment when it's hot because the body can perspire and control its temperature more effectively when evaporation is rapid. The rate of evaporation decreases as the amount of water vapor in the air increases. When the body gets too warm, the blood vessels at the skin dilate so as to bring more blood to the surface which makes the skin red.

•13.11

The ice forms first where it is in contact with the tray (which is a better conductor than the air) and then across the top, thereby encapsulating some water. Since ice is not a very good conductor, the remaining water takes a while to freeze. The warm water will be cooled more effectively, especially by evaporation, and enough of it may evaporate so that it will win the race. Even so, there are a lot of variables and the cool water often freezes first.

•13.13

The red arrows indicate the rate of flow of heat that decreases in the exposed rod where there are losses (mostly via convection), and remains constant in the insulated rod where there are not. In the exposed rod, there is less and less thermal energy available and the flow diminishes. Correspondingly, the slope of the T-d curve, which is the *temperature gradient*, also diminishes. In the insulated rod, the heat flow is uniform and the temperature gradient constant; that is, the slope is constant. Heat is transported by the temperature gradient much as a liquid is propelled through a pipe by a potential energy gradient.

•13.17

Wearing the hair inside will trap air that will not be disturbed by external winds. The low thermal conductivity of the air layer is what is important and that is better achieved with the fur inside. To stay cool, wear light-colored, porous, light-weight clothes. These will reflect radiant energy and facilitate evaporation of sweat. If there isn't much water to be had, evaporation of perspiration should be restrained as much as possible.

•13.19

Thermal radiation from the Sun will strike a portion of your suit and it will absorb energy at a rate dependent upon its surrounding face characteristics. The intensity of that incident radiation is determined in part by the distance to the source the farther away, the less power per unit area. Moreover, the suit will radiate over its entire surface at a rate dependent on its temperature (which, in turn, is partly determined by the body's output). The same sort of thing will happen to the thermometer. It will read an equilibrium temperature that is determined, among other things, by the fraction of its area illuminated. At the distance of the Earth from the Sun, a sphere in thermal equilibrium will stabilize at about 290 K, not far from room temperature. Thus, the thermometer reads its own temperature and not that of space.

Answers to Odd-Numbered Multiple Choice Questions

1. e **3.** c **5.** c **7.** b **9.** a **11.** c **13.** a
15. b **17.** b **19.** c

Solutions to Selected Problems

13.5

Use Eq. (13.1), $Q = cm\,\Delta T$, to find ΔT, the change in temperature for water (of mass m) as it absorbs an amount of heat Q. Here $Q = 500\,\text{kcal}$, $c = c_\text{W} = 1.000\,\text{kcal/kg·K}$ is the specific heat capacity of water, and m can be found from the volume V_W and density ρ_W of water to be $m = \rho_\text{W} V_\text{W} = (1.00 \times 10^3\,\text{kg/m}^3)(30 \times 10^{-3}\,\text{m}^3) = 30\,\text{kg}$. Solve for ΔT:

$$\Delta T = \frac{Q}{c_\text{W} m} = \frac{500\,\text{kcal}}{(1.000\,\text{kcal/kg·K})(30\,\text{kg})} = 17\,\text{K} = 17\,\text{C}^\circ\,.$$

13.29

The heat Q_I released by the iron as it cools down is equal to the heat Q_W absorbed by the water as it warms up. Use Eq. (13.1) to find Q_I: $Q_\text{I} = c_\text{I} m_\text{I} \Delta T_\text{I}$. Here $c_\text{I} = 0.113\,\text{kcal/kg·K}$ is the specific heat capacity of iron, $m_\text{I} = 100\,\text{g} = 0.100\,\text{kg}$, and $\Delta T_\text{I} = 80℃ - 30℃ = 50\,\text{C}^\circ = 50\,\text{K}$. Equate this to the heat absorbed by the water, $Q_\text{W} = c_\text{W} m_\text{W} \Delta T_\text{W}$, where $c_\text{W} = 1.00\,\text{kcal/kg·K}$ and $\Delta T_\text{W} = 30℃ - 25℃ = 5\,\text{C}^\circ = 5\,\text{K}$, to obtain

$$Q_\text{I} = c_\text{I} m_\text{I} \Delta T_\text{I} = Q_\text{W} = c_\text{W} m_\text{W} \Delta T_\text{W}\,,$$

which we solve for m_W, the mass of the water:

$$m_\text{W} = \frac{c_\text{I} m_\text{I} \Delta T_\text{I}}{c_\text{W} \Delta T_\text{W}} = \frac{(0.113\,\text{kcal/kg·K})(0.100\,\text{kg})(50\,\text{K})}{(1.00\,\text{kcal/kg·K})(5\,\text{K})} = 0.11\,\text{kg}\,.$$

13.33

The initial KE of the bullet, KE_i, is converted into the thermal energy needed to raise the temperature of both the wooden block (W) and the lead bullet (L). Assuming that the bullet's initial temperature is the same as that of the block, then since they must reach the same final temperature the temperature change ΔT for the bullet and the block must also be the same.

The heat it takes for the temperature change to occur is $c_W m_W \Delta T + c_L m_L \Delta T$, which we equate to $\mathrm{KE_i}$ to obtain

$$\mathrm{KE_i} = c_W m_W \Delta T + c_L m_L \Delta T\,.$$

Note that $\mathrm{KE_i} = (540\,\mathrm{ft\cdot lb})(0.3048\,\mathrm{m/ft})(4.448\,\mathrm{N/lb}) = 730.2\,\mathrm{N\cdot m} = 730.2\,\mathrm{J}$, and $m_L = (158\,\mathrm{grains})(0.0648\,\mathrm{g/grain}) = 10.24\,\mathrm{g} = 0.01024\,\mathrm{kg}$; so the equation above gives

$$\begin{aligned}\Delta T &= \frac{\mathrm{KE_i}}{c_W m_W + c_L m_L} \\ &= \frac{730.2\,\mathrm{J}}{(1700\,\mathrm{J/kg\cdot K})(1.0\,\mathrm{kg}) + (130\,\mathrm{J/kg\cdot K})(0.01024\,\mathrm{kg})} \\ &= +0.43\,\mathrm{K} = +0.43\,\mathrm{C^\circ}\,.\end{aligned}$$

13.37

Suppose that the burned food releases an amount of heat Q, which is absorbed by the water (W) plus the aluminum (A). The heat absorbed by the water is given by Eq. (13.1) to be

$$Q_1 = c_W m_W (T_f - T_i)\,,$$

where $c_W = 4.186\,\mathrm{kJ/kg{\cdot}K}$ is the specific heat of the water, $m_W = 2.00\,\mathrm{kg}$ is its mass, $T_i = 20^\circ\mathrm{C}$ is the initial temperature of the water, and T_f its final temperature. Meanwhile the aluminum chamber and the cup, with a total mass of $m_A = 0.50\,\mathrm{kg} + 0.60\,\mathrm{kg} = 1.10\,\mathrm{kg}$, absorb a combined amount of heat of

$$Q_2 = c_A m_A (T_f - T_i)\,,$$

where $c_A = 0.900\,\mathrm{kJ/kg\cdot K}$. Note that the aluminum chamber and the cup must undergo the same temperature change as the water since they remain in thermal contact throughout the process. Thus

$$\begin{aligned}Q = Q_1 + Q_2 &= c_W m_W (T_f - T_i) + c_A m_A (T_f - T_i) \\ &= [(4.186\,\mathrm{kJ/kg\cdot K})(2.00\,\mathrm{kg}) + (0.900\,\mathrm{kJ/kg\cdot K})(1.10\,\mathrm{kg})]\,(32^\circ\mathrm{C} - 20^\circ\mathrm{C}) \\ &= 1.1 \times 10^2\,\mathrm{kJ}\,,\end{aligned}$$

which is equivalent to 27 kcal.

13.45

The heat it takes to raise the temperature of a body of water of mass m_W from T_i to T_f is given by $Q = c_W m_W (T_f - T_i)$, where $c_W = 4.186\,\mathrm{kJ/kg{\cdot}K}$ is the specific heat capacity of water. In this problem the heat comes from burning the hard coal, which has a heat of combustion of 33 MJ/kg. So if the mass of the coal to be burned is m, then the heat released is $Q = (33\,\mathrm{MJ/kg})m$, which should be equal to the heat needed for the given temperature change in the water:

$$Q = (33\,\mathrm{MJ/kg})m = c_W m_W (T_f - T_i)\,.$$

Plug in $m_{\mathrm{W}} = 1.00\,\mathrm{kg}$, $T_{\mathrm{i}} = 0℃$, and $T_{\mathrm{f}} = 100℃$ and solve for m:

$$m = \frac{c_{\mathrm{W}} m_{\mathrm{W}} (T_{\mathrm{f}} - T_{\mathrm{i}})}{33\,\mathrm{MJ/kg}} = \frac{(4.186\,\mathrm{kJ/kg\cdot K})(1.00\,\mathrm{kg})(100℃ - 0℃)}{33 \times 10^3\,\mathrm{kJ/kg}} = 0.013\,\mathrm{kg} = 13\,\mathrm{g}\,.$$

13.83

When water at above 0℃ is mixed with ice, whose initial temperature is 0℃, the water will release heat which is absorbed by the ice. Suppose that the heat released by the water is Q_1 when its temperature drops to 0℃, and denote the heat absorbed by the ice as Q_2 when all of it just melts. Then in general three possibilities exist:

(i) if $Q_1 > Q_2$, then there is more than enough thermal energy available from the water to cause all of the ice to melt and the final state will be 100% water, at a final temperature T_{f} higher than 0℃;

(ii) if $Q_1 = Q_2$, then there is just enough energy for all of the ice to melt so the final state is 100% water at $T_{\mathrm{f}} = 0℃$; and

(iii) if $Q_1 < Q_2$, then there is not enough energy to melt all the ice so the final state is a mixture of ice and water, at $T_{\mathrm{f}} = 0℃$.

According to the problem statement the mass of the water is equal to that of the ice to begin with. Call this mass m. Then in our case

$$Q_1 = -c_{\mathrm{W}} m \Delta T_{\mathrm{W}} = -(1.00\,\mathrm{kcal/kg\cdot K}) m (0℃ - 80℃) = m(80\,\mathrm{kcal/kg})$$

and

$$Q_2 = m L_{\mathrm{f}} = m(80\,\mathrm{kcal/kg})\,.$$

So to two significant figures $Q_1 = Q_2$, meaning that the final state will be 100% water (with a total mass of $2m$) at $T_{\mathrm{f}} = 0℃$ — this is case (ii) above.

13.85

Apply Eq. (13.1) to find the heat Q it takes for the water of mass m_{W} to undergo a temperature change ΔT_{W}:

$$Q = c_{\mathrm{W}} m_{\mathrm{W}} \Delta T_{\mathrm{W}} = (4.186\,\mathrm{kJ/kg\cdot K})(0.900\,\mathrm{kg})(100℃ - 5.0℃) = 358\,\mathrm{kJ} \approx 0.36\,\mathrm{MJ}\,.$$

If this much heat is absorbed by $m_{\mathrm{S}} = 1.20\,\mathrm{kg}$ of silver (S), which we assume undergoes no subsequent phase change, then the temperature change for the silver satisfies $Q = c_{\mathrm{S}} m_{\mathrm{S}} \Delta T_{\mathrm{S}}$, or

$$\Delta T_{\mathrm{S}} = \frac{Q}{c_{\mathrm{S}} m_{\mathrm{S}}} = \frac{358\,\mathrm{kJ}}{(0.230\,\mathrm{kJ/kg\cdot K})(1.20\,\mathrm{kg})} = 1297\,\mathrm{K} = 1297\,\mathrm{C}^\circ\,,$$

which means that the final temperature of the silver would be $5.0℃ + 1297\,\mathrm{C}^\circ = 1302℃$, which would have exceeded 960.8℃, the melting point of silver. This tells us that our assumption of no phase change for the silver is invalid, and so the silver must first reach 960.8°C and then at least partially liquefy as a result of the heat it absorbs. To reach 960.8°C the silver must absorb

$$Q_1 = c_s m_s \Delta T_s = (0.230\,\mathrm{kJ/kg{\cdot}K})(1.20\,\mathrm{kg})(960.8^\circ\mathrm{C} - 5.0^\circ\mathrm{C}) = 264\,\mathrm{kJ}\,,$$

leaving $Q - Q_1 = 358\,\mathrm{kJ} - 264\,\mathrm{kJ} = 94\,\mathrm{kJ}$ available for melting. The mass m'_s of silver that can be melted with this much heat is $m'_s = 94\,\mathrm{kJ}/L_f$, where $L_f = 109\,\mathrm{kJ/kg}$ is the heat of fusion for silver. This gives $m'_s = 0.86\,\mathrm{kg}$, meaning that the final state consists of 0.86 kg of liquid silver plus $1.20\,\mathrm{kg} - 0.86\,\mathrm{kg} = 0.34\,\mathrm{kg}$ of solid silver, at $T_f = 960.8^\circ\mathrm{C}$, the melting point of silver.

14 Thermodynamics

Answers to Selected Discussion Questions

•14.1

Air is drawn from A-B as the piston moves out. From B-C, it is compressed adiabatically and rises in temperature. Fuel is sprayed in at C and immediately explodes because of the high temperature. C-D is the isobaric combustion leg when heat enters. At D, the fuel is burnt and the piston continues to move outward. D-E is the rapid and therefore adiabatic expansion where the hot gas does work. E is the end of the cycle and it's there that the exhaust valve opens. E-B is an isovolumic drop to the outside pressure, accompanied by the loss of heat via the exhausting of hot gas. With the exhaust valve still open, the remaining gas is ejected as the piston moves inward from B to A.

•14.3

Steam (blue arrows) enters behind the piston and drives it to the left as the slide valve moves to the right. Steam in the left half of the cylinder (red arrows) escapes through the exhaust port to the low-pressure condenser. The valve on the right then closes, the steam expands, and the piston moves left. Now the valve on the right opens to the exhaust, steam enters via the valve on the left and the piston moves right.

•14.5

Warm, moist low-density air rises (via thermals) doing work on the atmosphere as it expands. Accordingly, since the process occurs quickly, and gases are very poor conductors, it will be essentially adiabatic. The internal energy of the uprush of air will decrease and its temperature will drop. Water will then condense out, forming a cloud.

•14.7

The Sun provides the energy stored chemically in wood, coal, and oil. Order increases locally as the plants and animals storing solar energy grow, ultimately to form fossil fuels, but increase in the disorder of the Sun overbalances that.

•14.9

The Sun evaporates water that comes down to the rivers and reservoirs as rain, thereafter to pour down a waterfall and drive a turbine. In effect, the Sun does work against gravity.

•14.11

The process is adiabatic and so isentropic. Work is done by the gas on the piston. Therefore, the internal energy decreases, and the temperature (T) drops. The volume obviously doubles and the pressure decreases. The entropy must stay constant since $Q = 0$.

•14.13

The engine actually exhausts gas at a much higher temperature than 300 K. Fuel is incompletely burned, heat is conducted and radiated from the engine, and friction is present as well.

•14.15

Yes it is possible, but with so many molecules the likelihood of all of them being in one corner is minute. Of course, if there were only 1 molecule flying around, the chance of "all the molecules" being in one corner would be fairly high. It's less likely with 10 molecules and much less still with 10^{28}.

Answers to Odd-Numbered Multiple Choice Questions

1. b **3.** d **5.** d **7.** b **9.** e **11.** b **13.** c
15. a **17.** d **19.** b

Solutions to Selected Problems

14.7

The change in internal energy, ΔU, is related to the heat Q that enters the system and the work W performed by the system via Eq. (14.2): $\Delta U = Q - W$. In this case $Q = 100\,\text{cal} = (100\,\text{cal})(4.186\,\text{J/cal}) = 418.6\,\text{J}$ and $W = -100.4\,\text{N}\cdot\text{m} = -100.4\,\text{J}$. (Note that $W < 0$ since the work is done on, rather than by, the system). Thus

$$\Delta U = Q - W = (+418.6\,\text{J}) - (-100.4\,\text{J}) = +519\,\text{J}\,.$$

14.23

The cylinder is thermally insulated so no heat flows in or out of it: $Q = 0$. Thus according to Eq. (14.2) the change in internal energy of the water-steam mixture inside the cylinder, ΔU, is related to the work W done by the steam as

$$\Delta U = Q - W = -W = -P\Delta V\,,$$

where P ($= 99\,\text{kPa}$) is the pressure the piston is pushing against and $\Delta V = +4950\,\text{cm}^3 = +4.950 \times 10^{-3}\,\text{m}^3$.

Now, the temperature inside the cylinder remains at 100°C so there is no change in the average molecular KE of the system, and so ΔU is solely due to change in molecular PE as a result of the phase change as the steam condenses into water:

$$\Delta U = -m_{\text{s}} L_{\text{v}}\,,$$

where m_{s} is the mass of the steam which condenses into water in the process and $L_{\text{v}} = 2259\,\text{kJ/kg}$ is the heat of vaporization for water. Here we noted that $\Delta U < 0$ as steam loses thermal energy while condensing into water.

Equate the two expressions obtained above for ΔU to obtain

$$\Delta U = -P\Delta V = -m_{\text{s}} L_{\text{v}}\,,$$

which gives

$$m_{\text{s}} = \frac{P\Delta V}{L_{\text{v}}} = \frac{(99 \times 10^3\,\text{Pa})(4.950 \times 10^{-3}\,\text{m}^3)}{2259 \times 10^3\,\text{J/kg}} = 2.2 \times 10^{-4}\,\text{kg} = 0.22\,\text{g}\,.$$

14.41

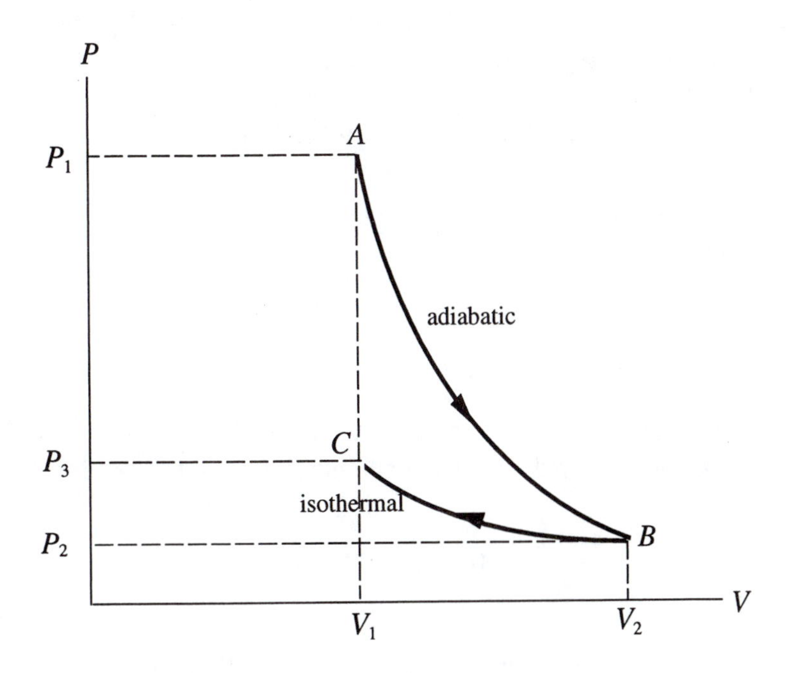

The P-V diagram depicting the process for the ideal gas in question is shown above. The gas starts from state A, where the pressure is $P_1 = 1.00\,\text{atm}$ and $V_1 = 22.4\,\text{liters}$; proceeding adiabatically to state B where its volume gets doubled: $V_2 = 2V_1$, before reaching state C isothermally with $V_3 = V_1 = 22.4\,\text{liters}$.

(a) During the adiabatic expansion from A to B the pressure of the system decreases from P_1 to P_2. As the gas gets compressed isothermally from B to C its pressure increases from P_2 to P_3. So clearly the lowest pressure the system attains is P_2. Since process $A \to B$ is adiabatic $P_1V_1^\gamma = P_2V_2^\gamma$, so

$$P_2 = P_1\left(\frac{V_1}{V_2}\right)^\gamma = P_1\left(\frac{V_1}{2V_1}\right)^{1.40} = (1.00\,\text{atm})\left(\frac{1}{2}\right)^{1.40} = 0.379\,\text{atm}\,,$$

which is equivalent to $(0.379\,\text{atm})(0.101\,3\,\text{MPa/atm}) = 0.038\,4\,\text{MPa}$.

(b) As the gas expands adiabatically from A to B its temperature drops from T_1 $(= 273\,\text{K})$ to T_2, its value in state B. The temperature of the system remains at T_2 throughout the rest of the process since $B \to C$ is isothermal. So the lowest temperature T_2 is first attained in state B. Apply the Ideal Gas Law to states A and B: $P_1V_1/T_1 = P_2V_2/T_2$, which we solve for T_2:

$$T_2 = T_1\left(\frac{P_2}{P_1}\right)\left(\frac{V_2}{V_1}\right) = (273\,\text{K})\left(\frac{0.379\,\text{atm}}{1.00\,\text{atm}}\right)\left(\frac{2V_i}{V_i}\right) = 207\,\text{K}\,.$$

Here we noted that 1.00 mol of any ideal gas occupies 22.4 liters at STP, so the initial temperature of the gas must be $T_1 = 273\,\text{K}$ $(0°\text{C})$.

14.53

Imagine a heat engine which operates between a high-temperature reservoir at T_H and a low-temperature reservoir at T_L. The maximum possible efficiency of such an engine is that of a Carnot engine, given by Eq. (14.13): $e_c = 1 - T_L/T_H$. In this case $T_H = 98.6°F = [(5/9)(98.6 - 32) + 273]\,K = 310\,K$ and $T_C = 20°C = (20 + 273)K = 293\,K$, so the maximum efficiency is

$$e_c = 1 - \frac{T_L}{T_H} = 1 - \frac{293\,K}{310\,K} = 0.055 = 5.5\%\,.$$

14.69

During each cycle of its operation the engine in question takes in an amount of heat $Q_H = 1200\,J$ and dumps $Q_L = 450\,J$ to a cool reservoir. From Eq. (14.12) the efficiency of the engine is

$$e = 1 - \frac{Q_L}{Q_H} = 1 - \frac{450\,J}{1200\,J} = 0.625 = 62.5\%\,.$$

Now consider the engine operating backwards. During each cycle it takes out an amount of heat Q_L from the low-temperature reservoir, receives an amount of work W_i from outside, and dumps an amount of heat Q_H to the high-temperature reservoir. Conservation of energy requires that $Q_L + W_i = Q_H$, so $Q_L = Q_H - W_i = 1200\,J - 1000\,J = 200\,J$. This tells us that, while Q_H is unchanged (at 1200 J), the value of Q_L is different (450 J for forward operation and 200 J for backward operation). Thus the engine cannot be reversible. (And consequently its efficiency must be lower than that of a Carnot engine, which is reversible.)

14.71

Suppose that the refrigerator pumps out an amount of heat Q_L from its interior at a temperature T_L while it is being supplied with an amount of work W_i. Then according to Eq. (14.17) at its very best the coefficient of performance of such a refrigerator can attain is $\eta_c = Q_L/W_i = T_L/(T_H - T_L)$, where T_H is the temperature of the environment into which the heat from the refrigerator is dumped.

Solve for W_i from the expression above: $W_i = Q_L(T_H - T_L)/T_L$. Here Q_L is the heat that must be removed from 2000 lb of water to freeze it into ice: $Q_L = mL_f$, where $m = (2000\,lb)(0.4536\,kg/lb) = 907.2\,kg$ and $L_f = 333.7\,kJ/kg$ (the heat of fusion of ice). Also, $T_H = 88°F = [\frac{5}{9}(88 - 32) + 273]K = 304\,K$ and $T_L = 32°F = [\frac{5}{9}(32 - 32) + 273]K = 273\,K$. Substitute these numerical values into the expression for W_i obtained above to find

$$\begin{aligned} W_i &= Q_L\left(\frac{T_H - T_L}{T_L}\right) = (mL_f)\left(\frac{T_H - T_L}{T_L}\right) \\ &= (907.2\,kg)(333.7\,kJ/kg)\left(\frac{304\,K - 273\,K}{273\,K}\right) \\ &= 3.45 \times 10^7\,J\,. \end{aligned}$$

Since this much work is to be delivered in $t = 1\,\mathrm{d} = 86\,400\,\mathrm{s}$, the corresponding power that must be supplied to the refrigerator is

$$P = \frac{W_i}{t} = \frac{3.45 \times 10^7\,\mathrm{J}}{86\,400\,\mathrm{s}} = 399\,\mathrm{W}\,.$$

14.75

First, find the work input W_i needed for an ideal refrigerator to expel an amount of heat Q_L from the cold chamber. Suppose that the refrigerator pumps out an amount of heat Q_L from its interior at a temperature T_L while it is being supplied with an amount of work W_i. Then according to Eq. (14.17) at its very best the coefficient of performance of such a refrigerator can attain is $\eta_c = Q_L/W_i = T_L/(T_H - T_L)$, where T_H is the temperature of the environment into which the heat from the refrigerator is dumped. Solve for W_i from the expression above:

$$W_i = Q_L \frac{T_H - T_L}{T_L}\,.$$

In the present case $T_H = 27℃ = (27 + 273)\mathrm{K} = 300\,\mathrm{K}$ is the temperature of the environment and $T_L = -4.0℃ = (-4.0 + 273)\mathrm{K} = 269\,\mathrm{K}$ is that of the cold chamber. Now consider a time interval $t = 1.0\,\mathrm{min} = 60\,\mathrm{s}$, during which the heat removed from the cold chamber is $Q_L = 360\,\mathrm{J}$. Divide both sides of the equation for W_i by t to find the electric power P needed:

$$P = \frac{W_i}{t} = \left(\frac{Q_L}{t}\right)\left(\frac{T_H - T_L}{T_L}\right) = \left(\frac{360\,\mathrm{J}}{60\,\mathrm{s}}\right)\left(\frac{300\,\mathrm{K} - 269\,\mathrm{K}}{269\,\mathrm{K}}\right) = 0.69\,\mathrm{W}\,.$$

14.81

A Carnot cycle consists of two isothermal processes at T_L and T_H, respectively, plus two adiabatic processes for which $\Delta S = 0$ (since $Q = 0$). In a T-S diagram, an isothermal process is represented by a straight line parallel to the T-axis, while an adiabatic process corresponds to a straight line parallel to the S-axis (since S does not change). Therefore in a T-S diagram a Carnot cycle assumes the shape of a rectangle, as shown below. Here $A \to B$ and $C \to D$ are isothermal, while $B \to C$ and $D \to A$ are adiabatic. The system takes in an amount of heat Q_H from A to B, and disposes of an amount of heat Q_L from C to D.

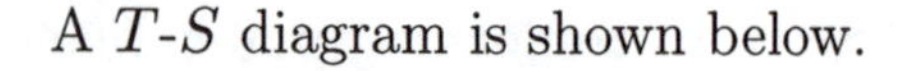

A T-S diagram is shown below.

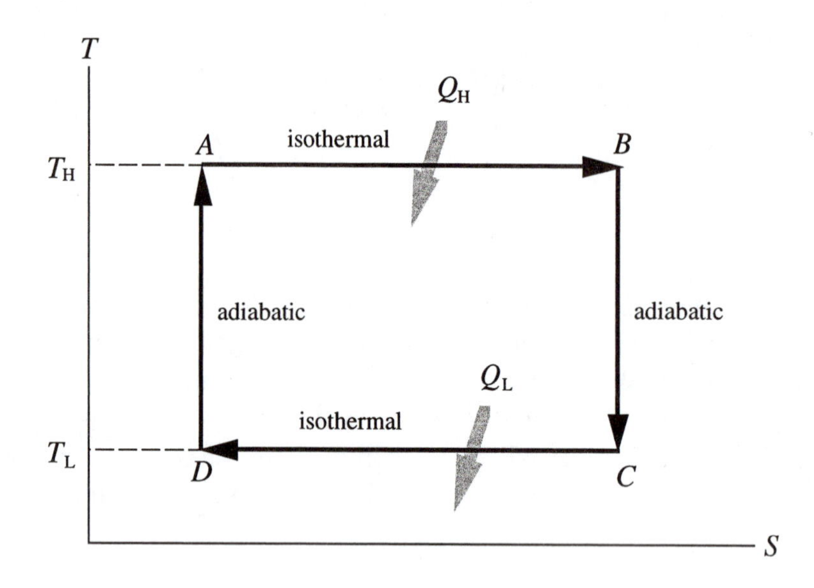

14.89

According to Eq. (14.19) the heat Q absorbed by a system at a constant temperature T is related to its entropy change ΔS by $\Delta S = Q/T$, or $Q = T\Delta S$. From Fig. P89 the temperature of the system is $T_{\rm H} = 300\,{\rm K}$ as its entropy changes from $10\,{\rm kJ/K}$ to $40\,{\rm kJ/K}$, so the heat intake is

$$Q_{\rm H} = T_{\rm H}\Delta S_{\rm H} = (300\,{\rm K})(30\,{\rm kJ/K}) = 9.0\times10^3\,{\rm kJ} = 9.0\,{\rm MJ}\,.$$

Similarly, as the system decreases its entropy by the same amount ($30\,{\rm kJ/K}$) at $T_{\rm L} = 100\,{\rm K}$, the heat that passes through the system is

$$Q_{\rm L} = T_{\rm L}\Delta S_{\rm L} = (100\,{\rm K})(-30\,{\rm kJ/K}) = -3.0\times10^3\,{\rm kJ} = -3.0\,{\rm MJ}\,,$$

where the minus sign indicates that heat is removed from, rather than added to, the system. The net work done by the system per cycle is $W_{\rm o} = Q_{\rm H} - |Q_{\rm L}| = 9.0\,{\rm MJ} - 3.0\,{\rm MJ} = 6.0\,{\rm MJ}$. The efficiency of the engine follows from Eq. (14.12):

$$e = \frac{W_{\rm o}}{Q_{\rm i}} = \frac{6.0\,{\rm MJ}}{9.0\,{\rm MJ}} = 0.67 = 67\%\,.$$

14.93

A common misconception associated with free expansion is that work has to be done as the gas expands. In fact, work is done by the gas only when the gas has to overcome some resistance, measured in terms of an external pressure which is against the expansion. In a *free* expansion, however, the gas does not have to push its way through — it is expanding into a vacuum, where there is no external pressure that would work against its expansion. So

$$W = 0 \qquad \text{(for a free expansion)}.$$

Also, since the expansion is adiabatic $Q = 0$. The First Law of Thermodynamics then requires that

$$\Delta U = Q - W = 0 \qquad \text{(for a free expansion)},$$

which implies that the temperature, which is proportional to the internal energy U for an ideal gas, does not change, either. Thus

$$T_f = T_i \qquad \text{(for a free expansion)}.$$

The free expansion is obviously an *irreversible* process. To find ΔS from $dS = dQ/T$ we must first identify a *reversible* path which leads from the initial state (P_i, V_i, T_i) to the final state (P_f, V_f, T_f), as $dS = dQ/T$ is valid only for a reversible process. Since $T_i = T_f$, one natural choice would be a reversible isothermal process leading from i to f, as shown in the diagram to the right.

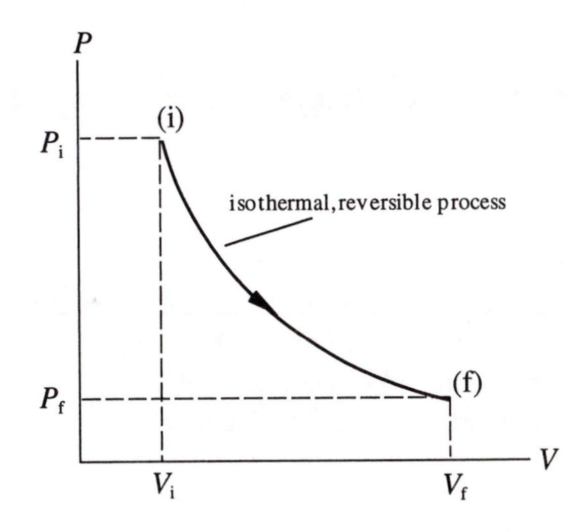

For the isothermal path $dU = 0$, and so

$$dS = \frac{dQ}{T} = \frac{dU + dW}{T} = \frac{dW}{T} = \frac{PdV}{T} = \frac{(nRT/V)\,dV}{T} = nR\frac{dV}{V}.$$

Thus

$$\Delta S = \int_{S_i}^{S_f} dS = nR\int_{V_i}^{V_f}\frac{dV}{V} = nR\Big[\ln V\Big]_{V_i}^{V_f} = nR\ln\frac{V_f}{V_i}.$$

Since $V_f > V_i$ as a result of the expansion, $\Delta S > 0$, which is expected since the free-expansion process is an adiabatic, spontaneous, irreversible one, which always leaves the entropy increased. (Note that $\ln x > 0$ for $x = V_f/V_i > 1$).

15 *Electrostatics: Forces*

Answers to Selected Discussion Questions

•15.1

The paper is electrically polarized by the field, and is therefore attracted to the charged conductor (or sheet of glass). It is only after a little while that charge is transferred to it from the highly charged object. At that point, the paper has a net charge with the same polarity as the conductor (or glass) and is repelled.

•15.3

The field is strongest in the region between the charges. The field lines are perpendicular to the walls and because the chamber is grounded there is no outer-surface charge and no field outside.

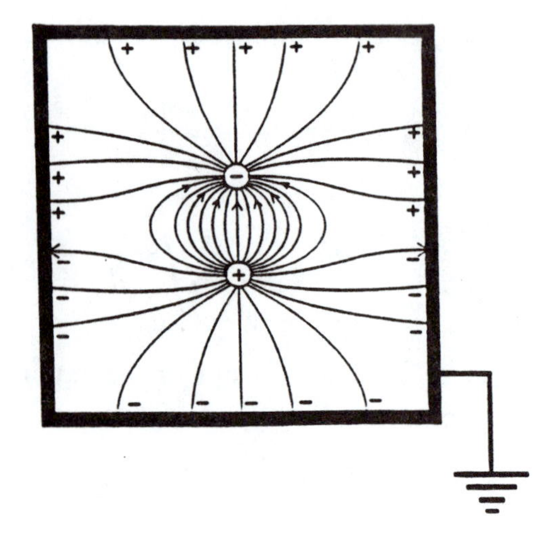

•15.7

Rubbing the tube electrified it and the free charges flowed along the moistened string, which was a fairly good conductor. The lead ball became charged as if it had been touched directly by the glass rod. It then polarized the leaf, which was attracted upward to it.

•15.9

As we will discuss later, the typewriter radiates an electromagnetic wave that the antenna wire picks up and superimposes on the signal coming down from the roof. One thing to do is to wrap the antenna wire, in the vicinity of the typewriter, with aluminum foil and ground it, thus shielding it. Another (though we will not discuss it until later), is to reorient the portion of the wire near the typewriter so it is not along the radiated E-field.

•15.15

The case of attraction is always ambiguous — the elephant might be either negative or neutral, whereupon the positive ball would induce a negative charge and be attracted. The repulsion unambiguously means that the elephant was negative.

Answers to Odd-Numbered Multiple Choice Questions

1. c **3.** b **5.** c **7.** a **9.** a **11.** d **13.** b
15. c **17.** d **19.** e

Solutions to Selected Problems

15.13

Refer to the figure to the right. It is clear from the symmetry of the configuration that the direction of the net force $\vec{\mathbf{F}}_3$ exerted by charges 1 and 2 on 3 is in the y-direction, so we need only to consider the y-component of the forces exerted by charges 1 and 2 on 3. The magnitude of the force from 1 is $F_{31} = k_0 q_1 q_3 / r_{13}^2$, and its y-component is

$$F_{31,y} = F_{31}\cos\theta = \left(k_0 \frac{q_1 q_3}{r_{13}^2}\right)\cos\theta\,;$$

and similarly the y-component of the force on charge 3 due to the other charge (charge 2) is $F_{32,y} = F_{32}\cos\theta = (k_0 q_2 q_3 / r_{23}^2)\cos\theta$. Since $q_1 = q_2 = q_3 = q$ and $r_{13} = r_{23} = r$, the two y-components are the same and the magnitude of the net force in the y-direction is

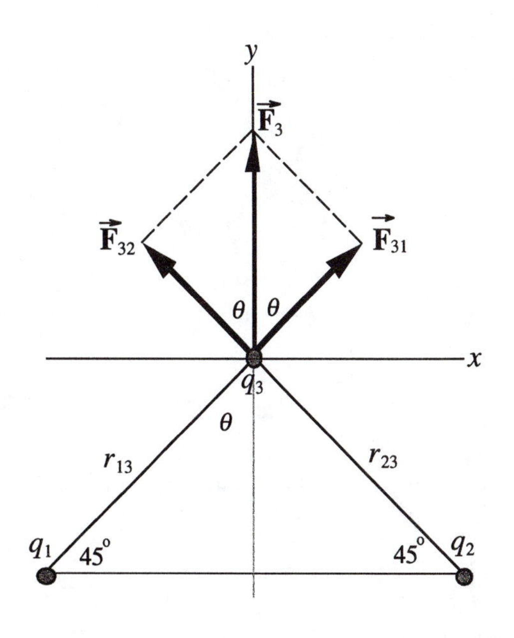

$$F_{3y} = F_{31,y} + F_{32,y} = k_0 \frac{q_1 q_3}{r_{12}^2} \cos\theta + k_0 \frac{q_2 q_3}{r_{13}^2} \cos\theta = 2k_0 \frac{q^2}{r^2} \cos\theta\,.$$

Plug in $q = +25\,\text{nC} = +2.5 \times 10^{-8}\,\text{C}$, $r = 1.0\,\text{m}$, and $\theta = 45°$ to obtain

$$\begin{aligned}\vec{\mathbf{F}}_3 &= F_{3y}\,\hat{\mathbf{j}} = 2k_0 \frac{q^2}{r^2} \cos\theta\,\hat{\mathbf{j}} \\ &= \frac{2(8.99 \times 10^9\,\text{N}\cdot\text{m}^2/\text{C}^2)(+2.5 \times 10^{-8}\,\text{C})^2(\cos 45°)}{(1.0\,\text{m})^2}\,\hat{\mathbf{j}} \\ &= (7.9 \times 10^{-6}\,\text{N})\,\hat{\mathbf{j}}\,.\end{aligned}$$

The force is in the positive y-direction, away from q_1 and q_2.

15.27

By Superposition Principle the net force exerted on q_1 by the other three charges is given by $\vec{\mathbf{F}}_1 = \vec{\mathbf{F}}_{12} + \vec{\mathbf{F}}_{13} + \vec{\mathbf{F}}_{14}$, where

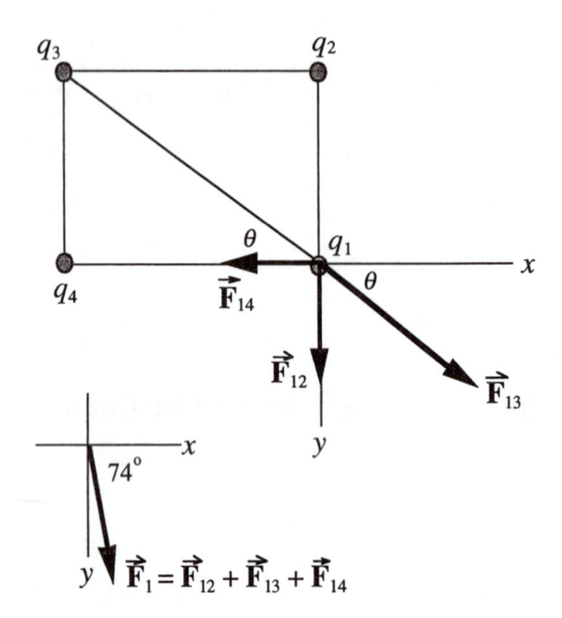

$$\begin{aligned}F_{12} &= k_0 \frac{q_1 q_2}{r_{12}^2} \\ &= \frac{(8.99 \times 10^9\,\text{N}\cdot\text{m}^2/\text{C}^2)(36\,\mu\text{C})(100\,\mu\text{C})}{(3.0\,\text{m})^2} \\ &= 3.596\,\text{N}\,,\end{aligned}$$

$$F_{13} = k_0 \frac{q_1 q_3}{r_{13}^2} = \frac{(8.99 \times 10^9\,\text{N}\cdot\text{m}^2/\text{C}^2)(125\,\mu\text{C})(100\,\mu\text{C})}{(3.0\,\text{m})^2 + (4.0\,\text{m})^2} = 4.495\,\text{N}\,,$$

and

$$F_{14} = k_0 \frac{q_1 q_4}{r_{14}^2} = \frac{(8.99 \times 10^9\,\text{N}\cdot\text{m}^2/\text{C}^2)(32\,\mu\text{C})(100\,\mu\text{C})}{(4.0\,\text{m})^2} = 1.798\,\text{N}\,.$$

Now add the x-components of these forces to obtain

$$F_{1x} = F_{12,x} + F_{13,x} + F_{14,x} = 0 + F_{13} \cos\theta - F_{14} = (4.495\,\text{N})\left(\frac{4.0\,\text{m}}{5.0\,\text{m}}\right) - 1.798\,\text{N} = 1.798\,\text{N}\,;$$

and add the y-components of these forces to obtain

$$F_{1y} = F_{12,y} + F_{13,y} + F_{14,y} = F_{12} + F_{13} \sin\theta + 0 = 3.596 + (4.495\,\text{N})\left(\frac{3.0\,\text{m}}{5.0\,\text{m}}\right) + 0 = 6.293\,\text{N}\,.$$

Thus the magnitude of $\vec{\mathbf{F}}_1$ is

$$F_1 = \sqrt{F_{1x}^{\,2} + F_{1y}^{\,2}} = \sqrt{(1.798\,\mathrm{N})^2 + (6.293\,\mathrm{N})^2} = 6.5\,\mathrm{N}\,,$$

and the angle ϕ between $\vec{\mathbf{F}}$ and the positive x-axis is

$$\phi = \tan^{-1}\left(\frac{F_{1y}}{F_{1x}}\right) = \tan^{-1}\left(\frac{6.293\,\mathrm{N}}{1.798\,\mathrm{N}}\right) = 74^\circ\,.$$

15.47

Draw an x-axis passing through the two charges in question, which are denoted as q_1 and q_2. The magnitudes of the electric fields $\vec{\mathbf{E}}_1$ and $\vec{\mathbf{E}}_2$ at the third vertex of the equilateral triangle due to the two charges are the same:

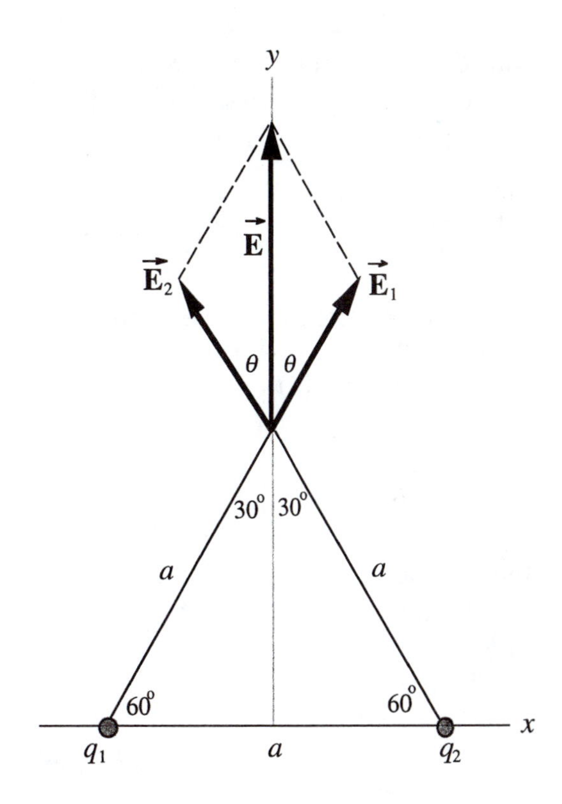

$$E_1 = k_0\frac{q_1}{a^2} = E_2 = k_0\frac{q_2}{a^2}\,,$$

where $q_1 = q_2 = +20\,\mu\mathrm{C}$ and $a = 2.0\,\mathrm{m}$. By symmetry the net $\vec{\mathbf{E}}$-field at that point, $\vec{\mathbf{E}} = \vec{\mathbf{E}}_1 + \vec{\mathbf{E}}_2$, must be in the positive y-direction:

$$\vec{\mathbf{E}} = (E_{1y} + E_{2y})\,\hat{\mathbf{j}} = k_0\frac{q_1}{a^2}\cos\theta\,\hat{\mathbf{j}} + k_0\frac{q_2}{a^2}\cos\theta\,\hat{\mathbf{j}}$$

$$= 2k_0\frac{q_1}{a^2}\cos\theta\,\hat{\mathbf{j}}$$

$$= \frac{2(8.99\times10^9\,\mathrm{N\cdot m^2/C^2})(+20\times10^{-6}\,\mathrm{C})(\cos 30^\circ)}{(2.0\,\mathrm{m})^2}\,\hat{\mathbf{j}}$$

$$= (7.8\times10^4\,\mathrm{N/C})\,\hat{\mathbf{j}}\,.$$

15.63

The electric force $\vec{\mathbf{F}}$ exerted by an electric field $\vec{\mathbf{E}}$ on an electron of charge $-q_e$ is given by Eq. (15.4) to be $\vec{\mathbf{F}} = -q_e\vec{\mathbf{E}}$. The resulting acceleration $\vec{\mathbf{a}}$ of the electron follows from Newton's Second Law as

$$\vec{\mathbf{a}} = \frac{\vec{\mathbf{F}}}{m_e} = \frac{-q_e\vec{\mathbf{E}}}{m_e}\,,$$

where m_e is its mass. The magnitude of the acceleration is then

$$a = \frac{q_e E}{m_e} = \frac{(1.602 \times 10^{-19}\,\text{C})(1.5 \times 10^4\,\text{N/C})}{9.109\,39 \times 10^{-31}\,\text{kg}} = 2.6 \times 10^{15}\,\text{m/s}^2\,.$$

Since the electron carries a negative charge the direction of $\vec{\mathbf{a}}$ is opposite to that of $\vec{\mathbf{E}}$.

15.73

In Example (15.7) we computed the electric field at point P due to a ring of radius R, carrying a uniformly distributed charge Q, to be $k_0 qx/(x^2 + R^2)^{3/2}$. In the present case we have a two-dimensional disc, which can be thought of as a collection of thin, concentric rings. Now consider one of these rings with radius r and thickness dr, as shown in Fig. P73. Since its area is $dA = 2\pi r\, dr$, the charge it carries is $dQ = \sigma dA = \sigma(2\pi r\, dr)$, and so its contribution to the electric field at point P must be

$$d\vec{\mathbf{E}}_x = dE_x\,\hat{\mathbf{i}} = \frac{k_0\, dQ\, x}{(x^2 + r^2)^{3/2}}\,\hat{\mathbf{i}} = \frac{k_0 (2\pi\sigma r\, dr)x}{(x^2 + r^2)^{3/2}}\,\hat{\mathbf{i}}\,,$$

which is clear following the result of Example (15.7), by replacing R with r and Q with $dQ = 2\pi\sigma r dr$. To sum over the contribution from all the rings making up the disc, integrate over r from $r = 0$ to $r = R$:

$$\vec{\mathbf{E}}_x = \int dE_x\,\hat{\mathbf{i}} = \pi k_0 \sigma x \int_0^R \frac{2r\, dr}{(x^2 + r^2)^{3/2}}\,\hat{\mathbf{i}}\,.$$

One of the ways to evaluate the last integral above is to introduce a new variable $u = x^2 + r^2$, which varies from x^2 to $x^2 + R^2$ as r itself varies from 0 to R. Note that $2r\, dr = d(r^2) = d(x^2 + r^2) = du$, and so

$$\int_0^R \frac{2r\, dr}{(x^2 + r^2)^{3/2}} = \int_{x^2}^{x^2+R^2} \frac{du}{u^{3/2}} = \left[-\frac{2}{u^{1/2}}\right]_{x^2}^{x^2+R^2} = \frac{2}{x} - \frac{2}{\sqrt{x^2 + R^2}}\,.$$

Hence

$$\vec{\mathbf{E}}_x = \pi k_0 \sigma x \left(\frac{2}{x} - \frac{2}{\sqrt{x^2 + R^2}}\right)\hat{\mathbf{i}} = 2\pi k_0 \sigma \left(1 - \frac{x}{\sqrt{x^2 + R^2}}\right)\hat{\mathbf{i}}\,.$$

15.89

Suppose that one plate is charged with Q and the other with $-Q$. Then the charge density of the two plates are $\sigma = Q/A$ and $\sigma' = -Q/A$, respectively, where $A = 100\,\text{cm} \times 50\,\text{cm} = 5.0 \times 10^3\,\text{cm}^2 = 0.50\,\text{m}^2$. According to Eq. (15.18) the electric field E between the two parallel plates is then

$$E = \frac{\sigma}{\varepsilon} = \frac{Q/A}{\varepsilon}\,.$$

Noting that $\varepsilon = \varepsilon_0$ in this case, we may solve for Q to obtain

$$Q = \varepsilon_0 AE = (8.8542 \times 10^{-12}\,\mathrm{C^2/N{\cdot}m^2})(0.50\,\mathrm{m^2})(1000\,\mathrm{N/C}) = 4.4\,\mathrm{nC}\,.$$

15.101

The charge distribution in this problem is spherically symmetrical, and so is the resulting electric field. To utilize this symmetry, draw a Gaussian surface in the shape of a sphere of radius r ($r < R$), concentric with the uniformly charged sphere. Since the $\vec{\mathbf{E}}$-field on the Gaussian surface is radial and uniform in magnitude the electric flux through the Gaussian surface is

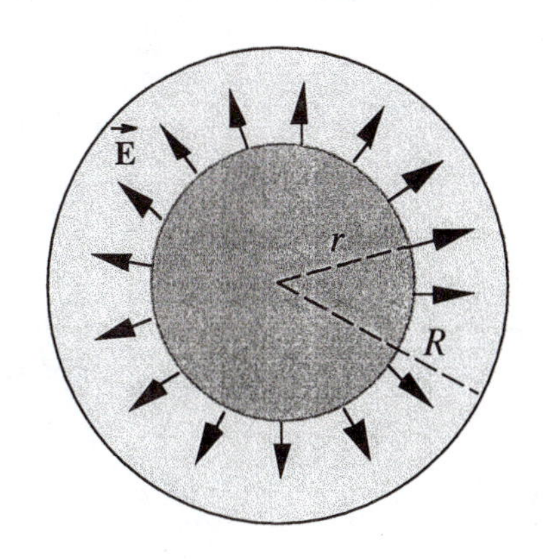

$$\oint E_\perp\, dA = \oint E\, dA = E \oint dA = EA = E(4\pi r^2)\,,$$

which we equate to $\sum q_\bullet/\varepsilon_0$ in accordance with Gauss's Law: $E(4\pi r^2) = \sum q_\bullet/\varepsilon_0$, where $\sum q_\bullet$ is the charge enclosed by the Gaussian surface. This gives

$$E = \frac{\sum q_\bullet}{4\pi\varepsilon_0 r^2}\,.$$

For uniform charge distribution $\sum q_\bullet = \rho V = (Q/V_0)V$, where $V = 4\pi r^3/3$ is the volume enclosed by the Gaussian surface and $V_0 = 4\pi R^3/3$ is the volume of the entire sphere. Thus $V/V_0 = r^3/R^3$, and so

$$\sum q_\bullet = Q\left(\frac{r^3}{R^3}\right)\,.$$

Plug this result into the expression for E, and introduce the unit vector $\hat{\mathbf{r}} = \vec{\mathbf{r}}/r$ to obtain

$$\vec{\mathbf{E}} = E\,\hat{\mathbf{r}} = \frac{\sum q_\bullet}{4\pi\varepsilon_0 r^2}\,\hat{\mathbf{r}} = \frac{Q(r^3/R^3)}{4\pi\varepsilon_0 r^2}\,\hat{\mathbf{r}} = k_0\frac{Q\vec{\mathbf{r}}}{R^3}\,.$$

The magnitude of the field increases linearly with the radius r, reaching $k_0 Q/R^2$ at $r = R$ (as expected). Also note that $E = 0$ at $r = 0$, which agrees with symmetry considerations.

16 Electrostatics: Energy

Answers to Selected Discussion Questions

•16.3

The neutral conductor becomes polarized with a negative induced charge near the original sphere and a positive charge on the far side. These charges, in turn, contribute to the potential, with the negative charges dominating and reducing the potential at the original sphere. Accordingly, the potential of the original sphere (that is, inside) is lower. Outside, it drops rapidly to a finite value that is constant across the neutral conductor.

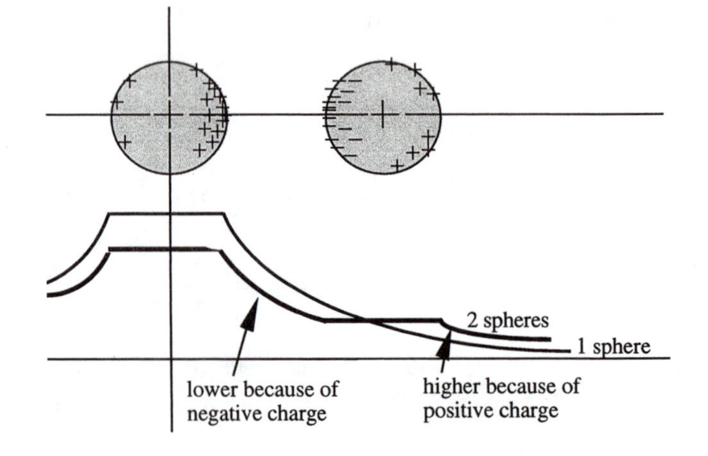

•16.5

Since no charge flows from or to the neutral conductor while touching the inside wall, the two must be at the same potential, which is uniform in the cavity, implying that the conductor assumes the potential of the field-free region. In a region where there is a field, the conductor will distort the field pattern and change the potential. If more charge is added to the outer conductor, the potential inside increases and the potential of the neutral body increases.

•16.9

The field lines go from the positive charges to the negative ones. There is no field at the very center. The equipotentials surround each of the charges, being negative in the vicinity of the negative charge. There are two zero equipotential planes that pass through the center point.

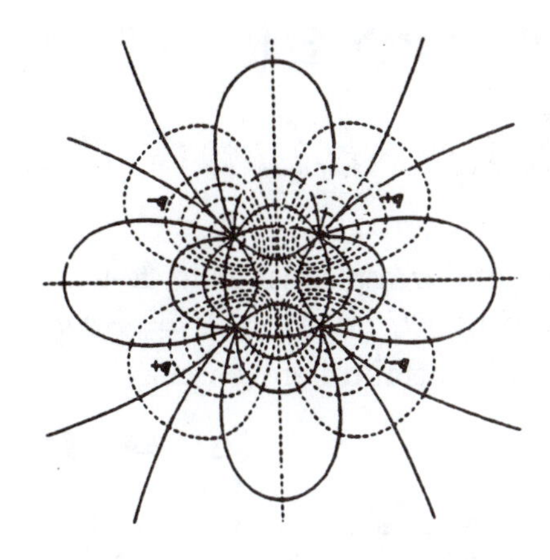

•16.11

Bringing a positive charge to a point in space introduces an outwardly directed field and a positive potential everywhere in the region; a negative charge does just the opposite, contributing a negative potential and so lowering the net potential everywhere. Imagine a gold-leaf electroscope attached to a positively charged plate. The electroscope reads the potential of the plate. Now bring a positive charge near the plate; positive charge from the plate will be repelled back to the scope, the leaves will part even more, and it will thereby indicate an increase in potential.

•16.15

There cannot be a charge since that would produce a field and therefore a potential gradient, but the potential is constant.

•16.19

The electric field lines are perpendicular to the grid of curved equipotentials. The field lines, which become fairly straight, converge to point P (from the anode on the right) and then diverge away from P (on the left). Electrons in the beam approaching from the left are consequently accelerated toward P.

Answers to odd-Numbered Multiple Choice Questions

1. b **3.** c **5.** b **7.** d **9.** d **11.** d **13.** c
15. d **17.** b **19.** b

Solutions to Selected Problems

16.9

The uniform electric field E which produces a potential difference of ΔV over a distance d is given by Eq. (16.6): $\Delta V = \pm Ed$. Plug in $E = 1.00\,\mathrm{V/m}$ and $d = 10.0\,\mathrm{cm} = 0.100\,\mathrm{m}$ to obtain the potential difference ΔV that needs to the applied:

$$\Delta V = \pm Ed = \pm(1.00\,\mathrm{V/m})(0.100\,\mathrm{m}) = \pm 0.100\,\mathrm{V}.$$

Here the $\pm$ sign means that either of the two plates can be the one that is at a higher potential than the other. Switching the sign of ΔV will only change the direction of the electric field (which is not specified in the problem statement), but not its magnitude.

16.11

According to the Principle of Conservation of Energy, the gain in KE for the electron is equal to the loss in its *electrical*-PE: $\Delta\mathrm{KE} = -\Delta\mathrm{PE_E} = -(-q_e)\Delta V$. Here $-q_e = -1.602 \times 10^{-19}\,\mathrm{C}$ is the charge of the electron and $\Delta V = +500\,\mathrm{V}$, so

$$\Delta\mathrm{KE} = -\Delta\mathrm{PE_E} = q_e \Delta V = q_e(+500\,\mathrm{V}) = +500(q_e \cdot \mathrm{V}) = +500\,\mathrm{eV},$$

where we used the definition of electron-volts: $1\,\mathrm{eV} = 1\,q_e \cdot \mathrm{V}$. Thus the KE of the electron increases by 500 eV, while its *electrical*-$\mathrm{PE_E}$ decreases by the same amount.

16.15

Suppose that the voltage difference across the plates is ΔV. This would result in an electric field of magnitude $E = \Delta V/d$, where d is the separation between the plates. The magnitude of the electric force $\vec{\mathbf{F}}_\mathrm{E}$ exerted by such a field on an electron of charge $-q_e$ is then $F_\mathrm{E} = q_e E = q_e \Delta V/d$. For the electron to be suspended in midair, its (downward) weight F_W must be balanced by an (upward) F_E so that the net force exerted on it vanishes:

$$+\!\uparrow\sum F = F_\mathrm{E} - F_\mathrm{W} = \frac{q_e \Delta V}{d} - m_e g = 0,$$

where m_e is the mass of the electron. Solve for ΔV:

$$\Delta V = \frac{m_e g d}{q_e} = \frac{(9.11 \times 10^{-31}\,\mathrm{kg})(9.81\,\mathrm{m/s^2})(0.100\,\mathrm{m})}{1.602 \times 10^{-19}\,\mathrm{C}} = 5.58 \times 10^{-12}\,\mathrm{V}.$$

16.19

The magnitude of the potential difference ΔV over a distance d along a uniform electric field $\vec{\mathbf{E}}$ is related to the E-field via Eq. (16.6): $|\Delta V| = Ed$. In our case $|\Delta V| = 85\,\mathrm{mV}$ and $d = 8\,\mathrm{nm}$, so

$$E = \frac{|\Delta V|}{d} = \frac{85 \times 10^{-3}\,\mathrm{V}}{8.0 \times 10^{-9}\,\mathrm{m}} = 1.1 \times 10^{7}\,\mathrm{V/m}.$$

16.67

To find the x- and y-components of the electric field using $E_x = -\partial V/\partial x$ and $E_y = -\partial V/\partial y$, we first need to rewrite V as a function of x and y, rather than r and θ. Note from the diagram to the right that $\cos\theta = x/r$ and $r = (x^2 + y^2)^{1/2}$, which gives

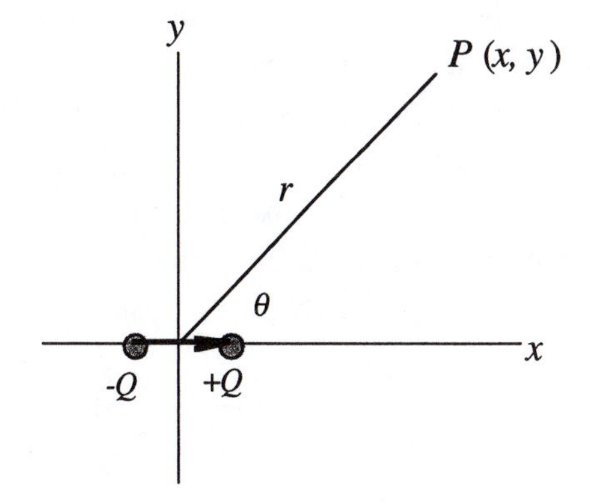

$$V = k_0 p\frac{\cos\theta}{r^2} = k_0 p\frac{x/r}{r^2} = \frac{k_0 px}{r^3} = \frac{k_0 px}{(x^2+y^2)^{3/2}}.$$

Thus

$$E_x = -\frac{\partial V}{\partial x} = -\frac{\partial}{\partial x}\left[\frac{k_0 px}{(x^2+y^2)^{3/2}}\right] = -k_0 p\left[\frac{1}{(x^2+y^2)^{3/2}} - \frac{3}{2}\frac{(2x)x}{(x^2+y^2)^{5/2}}\right]$$
$$= -\frac{k_0 p(y^2 - 2x^2)}{(x^2+y^2)^{5/2}};$$

and

$$E_y = -\frac{\partial V}{\partial y} = -\frac{\partial}{\partial y}\left[\frac{k_0 px}{(x^2+y^2)^{3/2}}\right] = -k_0 px\left[-\frac{3}{2}\frac{2y}{(x^2+y^2)^{5/2}}\right]$$
$$= -\frac{3k_0 pxy}{(x^2+y^2)^{5/2}}.$$

16.73

The reason why a warning sign is posted is because of the high voltage across the capacitor (say, 20 kV), which can be dangerous. If the tube voltage is V and the capacitance is C, then the charge Q stored in the CRT follows from Eq. (16.12) to be $Q = CV$. Plug in $V = 20\,\mathrm{kV} = 20 \times 10^3\,\mathrm{V}$ and $C = 500\,\mathrm{pF} = 200 \times 10^{-12}\,\mathrm{F}$ to obtain

$$Q = CV = (500 \times 10^{-12}\,\mathrm{F})(20 \times 10^3\,\mathrm{V}) = 1 \times 10^{-5}\,\mathrm{C},$$

which is a lot of charge.

16.85

The capacitance of a parallel-plate capacitor consisting of a pair of plates of area A separated by a gap of width d is given by Eq. (16.14): $C = \varepsilon A/d$. Here ε is the permittivity of the material in the gap. In this case $A = 100\,\mathrm{cm}^2$, $d = 1.0\,\mathrm{mm}$, and $\varepsilon = 10\varepsilon_0$; so

$$C = \frac{\varepsilon A}{d} = \frac{10(8.8542 \times 10^{-12}\,\mathrm{F/m})(100 \times 10^{-4}\,\mathrm{m}^2)}{1.0 \times 10^{-3}\,\mathrm{m}} = 8.9 \times 10^{-10}\,\mathrm{F}.$$

16.103

From the problem statement we know that there are 100 parallel-pate capacitors, each consisting of a sheet of paper sandwiched in between a pair of aluminum sheets. Suppose that the area of each aluminum sheet is A and the thickness of each paper sheet is d, then according to Eq. (16.14) the capacitance of each such capacitor is $C_1 = \varepsilon A/d$, where $\varepsilon = 4.1\varepsilon_0$ is the permittivity of the paper being used. All these capacitors are connected in parallel, since one sheet of aluminum from each capacitor is connected to a common wire while the other sheet is connected to another common wire. Thus from Eq. (16.15) the capacitance of the system is

$$C = 100\,C_1 = \frac{100\,\varepsilon A}{d} = \frac{100(4.1)(8.8542 \times 10^{-12}\,\mathrm{F/m})(0.12\,\mathrm{m} \times 0.50\,\mathrm{m})}{0.22 \times 10^{-3}\,\mathrm{m}} = 0.99\,\mu\mathrm{F}\,.$$

16.109

Each of the three capacitors (labeled C_1 through C_3) has one side connected to terminal A and the other to terminal B, so they are in parallel; and from Eq. (16.15) the equivalent capacitance is

$$C = C_1 + C_2 + C_3 = 2.0\,\mathrm{pF} + 3.0\,\mathrm{pF} + 4.0\,\mathrm{pF} = 9.0\,\mathrm{pF}\,.$$

16.117

The device is depicted in Fig. P117 is comprised of two parallel-plate capacitors, one consisting of the top and the middle plates and the other one the middle and the lower plates (note that the middle plate is shared by the two capacitors). The capacitance of each capacitor is $C_1 = C_2 = \varepsilon A/d$, where $A = 50\,\mathrm{cm} \times 60\,\mathrm{cm} = 0.30\,\mathrm{m}^2$, $d = 1.0\,\mathrm{mm}$, and $\varepsilon = 4.5\varepsilon_0$. The two capacitors are connected in parallel (since each capacitor has one plate connected to one common terminal and the other plate connected to the other common terminal). Thus the equivalent capacitance of the device is

$$C = C_1 + C_2 = \frac{2\varepsilon A}{d} = \frac{2(4.5)(8.8542 \times 10^{-12}\,\mathrm{F/m})(0.30\,\mathrm{m}^2)}{1.0 \times 10^{-3}\,\mathrm{m}} = 2.4 \times 10^{-8}\,\mathrm{F}\,.$$

16.129

Each of the four capacitors are connected directly across the two terminals of the battery (so they are in parallel). The equivalent capacitance C of all the four capacitors in parallel is $C = 1.0\,\mathrm{pF} + 8.0\,\mathrm{pF} + 9.0\,\mathrm{pF} + 12\,\mathrm{pF} = 30\,\mathrm{pF}$. It follows from Eq. (16.18) that, when a voltage difference of 12 V (supplied by the battery) is applied on it, the electrostatic energy stored in the circuit is

$$\mathrm{PE_E} = \frac{1}{2}CV^2 = \frac{1}{2}(30 \times 10^{-12}\,\mathrm{F})(12\,\mathrm{V})^2 = 2.2 \times 10^{-9}\,\mathrm{J} = 2.2\,\mathrm{nJ}\,.$$

17 Direct Current

Answers to Selected Discussion Questions

•17.1

The flow of heat along a metal rod is proportional to the temperature difference across its ends, just as the flow of charge is proportional to the voltage difference. In metals, free electrons serve to transport both electrical and thermal currents.

•17.5

This arrangement is for controlling a light bulb from two locations. Either switch will turn the bulb on or off independent of the other. As shown in Fig. Q5, the lamp is on. Throwing either switch open-circuits the bulb, shutting it off (as indicated here).

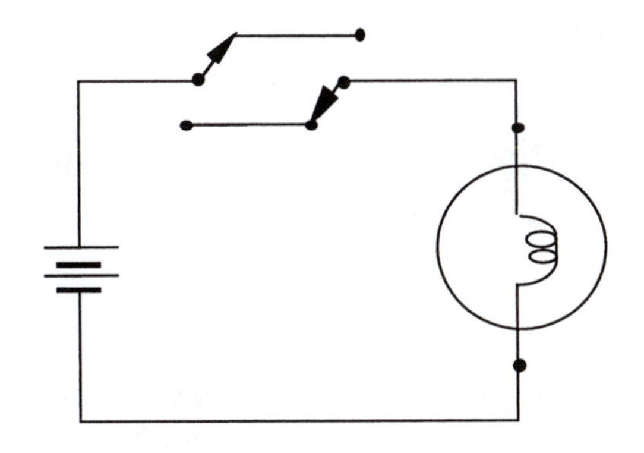

•17.7

The gasoline engine powers a separate electrical generator that operates all systems and sends its excess current to the battery to recharge it. Once the engine is running, the battery can be completely removed from the circuit. Leaving off all unnecessary electrical devices will allow more current to be provided to recharge the battery.

•17.11

The ammonium chloride dissociates into NH^+ (which, while current is circulating, migrates to the carbon rod) and Cl^-. Chlorine ions combine with the zinc electrode to make zinc chloride. If it were really dry, there would be no migration of the ions and it wouldn't work. Until only a few decades ago, the cell's outer casing was the thin zinc can, simply wrapped in a cardboard insulating sleeve. More often than not, the zinc would be eaten away and the ammonium chloride paste would leak out and corrode your flashlight. Even with today's steel casings, it's still not wise to store electronic equipment with batteries inside them for long periods.

•17.15

With current fanning out radially from the point of impact, there will be a voltage drop across the animal, and if its resistance is not very much greater than the ground's, a sizable current will pass through it. Squat.

Answers to Odd-Numbered Multiple Choice Questions

1. c **3.** d **5.** c **7.** e **9.** b **11.** b **13.** a
15. c **17.** d **19.** c **21.** a

Solutions to Selected Problems

17.5

The rate at which electric charge flows is the electric current. Suppose that an amount of charge Δq is drawn by the starter in a time interval Δt, then the average current flowing through the starter during that time interval is given by Eq. (17.1) as $I = \Delta q/\Delta t$. Plug in $I = 180\,\mathrm{A}$ and $\Delta t = 2.0\,\mathrm{s}$ and solve for Δq:

$$\Delta q = I\Delta t = (180\,\mathrm{A})(2.0\,\mathrm{s}) = 3.6 \times 10^2\,\mathrm{C}\,.$$

17.27

The current changes from its initial value of $I_i = 7.0\,\text{A}$ to its final value of $I_f = 3.0\,\text{A}$. Since the change in current is *linear*, its average value is $I_{av} = \frac{1}{2}(I_i + I_f) = \frac{1}{2}(7.0\,\text{A} + 3.0\,\text{A}) = 5.0\,\text{A}$. The total amount of charge Δq that passed through the battery during a time interval Δt is $\Delta q = I_{av}\Delta t$. Plug in $\Delta t = 6.0\,\text{h} = (6.0\,\text{h})(3600\,\text{s/h}) = 2.16 \times 10^4\,\text{s}$ to obtain

$$\Delta q = I_{av}\Delta t = (5.0\,\text{A})(2.16 \times 10^4\,\text{s}) = 1.1 \times 10^5\,\text{C} = 0.11\,\text{MC}.$$

17.49

The cross-sectional area A of a wire of diameter d is given by $A = \pi d^2/4$. If its length is L and its resistivity is ρ, then its resistance R is given by Eq. (17.6) as

$$R = \rho\frac{L}{A} = \rho\frac{L}{\pi d^2/4},$$

which we solve for d to obtain $d = \sqrt{4\rho L/\pi R}$. In this case $\rho = 1.7 \times 10^{-8}\,\Omega\cdot\text{m}$ for copper, $L = 1.0\,\text{mi} = 1.6 \times 10^3\,\text{m}$, and $R = 10\,\Omega$; so

$$d = \sqrt{\frac{4\rho L}{\pi R}} = \sqrt{\frac{4(1.7 \times 10^{-8}\,\Omega\cdot\text{m})(1.6 \times 10^3\,\text{m})}{\pi(10\,\Omega)}} = 1.9 \times 10^{-3}\,\text{m} = 1.9\,\text{mm}.$$

17.63

Suppose that the resistivity of a conductor is originally ρ_0. Then after the temperature changes by ΔT the new resistivity is given by Eq. (17.7) to be $\rho \approx \rho_0(1 + \alpha_0\Delta T)$, which corresponds to a change in resistivity of $\Delta\rho = \rho - \rho_0 \approx \rho_0(1 + \alpha_0\Delta T) - \rho_0 = \alpha_0\rho_0\Delta T$. Divide this expression for $\Delta\rho$ by ρ_0 to obtain

$$\frac{\Delta\rho}{\rho_0} \approx \frac{\alpha_0\rho_0\Delta T}{\rho_0} = \alpha_0\Delta T.$$

The assumption here is that the temperature coefficient remains constant over ΔT. Since $R = \rho L/A \propto \rho$, the fractional change in the resistance R of the filament is given by $\Delta R/R_0 = \Delta\rho/\rho_0 \approx \alpha_0\Delta T$. Using the value of α_0 at 20 °C from Table (17.3), $0.0045\,\text{K}^{-1}$, and noting that $\Delta T = 2800\,°\text{C} - 2200\,°\text{C} = 600\,\text{C}° = 600\,\text{K}$, we get

$$\frac{\Delta R}{R_0} \approx \alpha_0\Delta T = (0.0045\,\text{K}^{-1})(600\,\text{K}) = 2.7.$$

This approximation, however, is quite poor since it is clear from the data given in the problem statement that α_0 is not a constant in the given temperature range. From $T = 500\,°\text{C}$ to $1000\,°\text{C}$ the value of α_0 increases from $0.0057\,\text{K}^{-1}$ to $0.0089\,\text{K}^{-1}$. Assuming that α_0 continues to increase linearly with temperature at roughly the same rate, then from $T = 1000\,^\circ\mathrm{C}$ to $T = 2500\,^\circ\mathrm{C}$ the corresponding change in α_0 is roughly

$$\Delta\alpha_0 \approx \left(\frac{0.0089\,\mathrm{K}^{-1} - 0.0057\,\mathrm{K}^{-1}}{1000\,^\circ\mathrm{C} - 500\,^\circ\mathrm{C}}\right)(2500\,^\circ\mathrm{C} - 1000\,^\circ\mathrm{C}) \approx 0.01\,\mathrm{K}^{-1},$$

so at $T = 2500\,^\circ\mathrm{C}$ we have

$$\alpha_0 \approx \alpha_0\big|_{T=1000\,^\circ\mathrm{C}} + \Delta\alpha_0 = 0.0089\,\mathrm{K}^{-1} + 0.01\,\mathrm{K}^{-1} \approx 0.02\,\mathrm{K}^{-1}.$$

(Note that the nature of this rough estimation precludes us from keeping more than one significant figure in the final result.) Using the new data for α_0, which is measured at $T = 2500\,^\circ\mathrm{C}$, the medium temperature between $2200\,^\circ\mathrm{C}$ and $2800\,^\circ\mathrm{C}$, the fractional change in R can be re-estimated to be

$$\frac{\Delta R}{R_0} \approx \alpha_0 \Delta T = (0.02\,\mathrm{K}^{-1})(600\,\mathrm{K}) = 12,$$

which is considerably different from the previous estimation of 2.7.

17.91

The current I is the time rate at which charge q is delivered: $I = dq/dt$. Since $q = (4.00\,\mathrm{C/s^3})t^3 - (4.00\,\mathrm{C/s})t$,

$$\begin{aligned} I = \frac{dq}{dt} &= \frac{d}{dt}\left[(4.00\,\mathrm{C/s^3})t^3 - (4.00\,\mathrm{C/s})t\right] \\ &= (4.00\,\mathrm{C/s^3})(3t^2) - 4.00\,\mathrm{C/s} = (12.0\,\mathrm{C/s^3})t^2 - 4.00\,\mathrm{C/s} \\ &= (12.0\,\mathrm{A/s^2})t^2 - 4.00\,\mathrm{A}, \end{aligned}$$

where in the last step we noted the definition $1\,\mathrm{A} = 1\,\mathrm{C/s}$. The current density J is the current per unit cross-sectional area. Since the rod has a cross-sectional area of $A = 2.00\,\mathrm{mm}^2 = (2.00\,\mathrm{mm}^2)(10^{-3}\,\mathrm{m/mm})^2 = 2.00 \times 10^{-6}\,\mathrm{m}^2$, the current density at $t = 1.00\,\mathrm{s}$ is

$$J = \frac{I}{A} = \frac{(12.0\,\mathrm{A/s^2})t^2 - 4.00\,\mathrm{A}}{2.00 \times 10^{-6}\,\mathrm{m}^2}\bigg|_{t=1.00\,\mathrm{s}} = \frac{(12.0\,\mathrm{A/s^2})(1.00\,\mathrm{s})^2 - 4.00\,\mathrm{A}}{2.00 \times 10^{-6}\,\mathrm{m}^2} = 4.00\times10^6\,\mathrm{A/m^2}.$$

18 *Circuits*

Answers to Selected Discussion Questions

•18.3

This circuit was supposed to simply illustrate the Node Rule at G; 12 amps flow in, 12 amps flow out. But they overlooked something. A-H-G and B-H-G are short circuits; A and G, and B and G are all at the same potential. That means that no current flows through the resistors in branches A-G and B-G, and the diagram is incorrect as labeled.

•18.5

The switch is closed and the capacitor charges up slowly (depending on RC). As it does, the voltage across it increases until the breakdown voltage of the lamp is reached, at which point the capacitor rapidly discharges through the lamp, which flashes. The voltage across the capacitor, and therefore the lamp as well, drops rapidly only to slowly build up again for a repeat performance.

•18.7

The circuit first sees the edge of the pulse and a voltage of V. Charge and voltage both gradually build on C until the signal drops to 0 at which point they both decay. If the time constant is short enough, V_C will reach nearly 0 before the next pulse arrives. Since the drop across the capacitor is initially zero, V_R starts at a maximum and decays exponentially to 0.

When the input signal drops to 0, it's as if the input terminals were shorted. The voltage in the circuit comes from the charged capacitor, and the resistor is across its terminals. When the upper input circuit terminal was positive, the current circulated clockwise. When the input is shorted, current (discharged from the capacitor) circulates counterclockwise. Hence, the voltage polarity on the resistor is reversed and decays from there to zero as the capacitor discharges. In general, $V_R = V - V_C$.

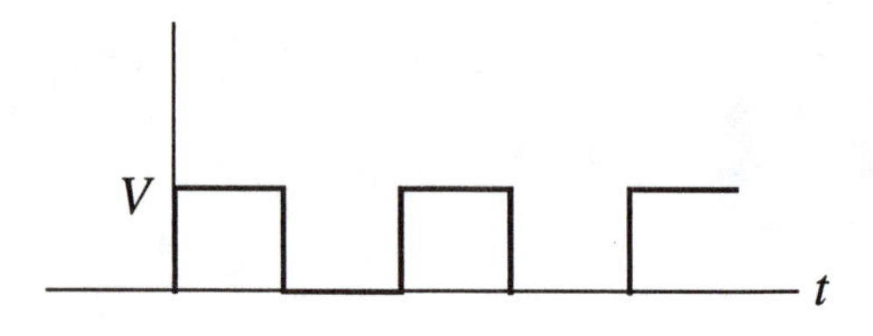

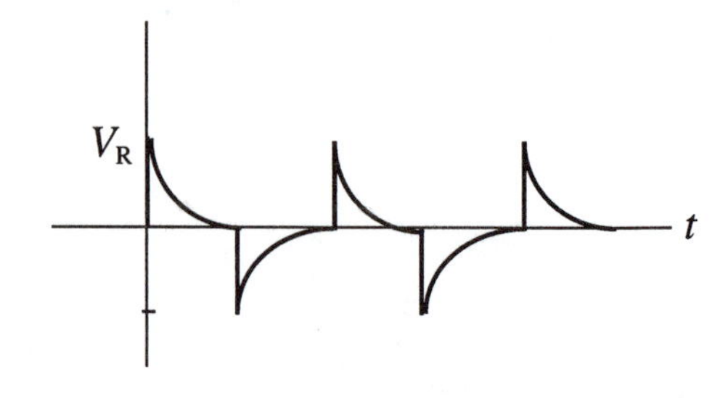

•18.9

The benchmark of direct current is that it does not change direction, which means that a dc voltage must not change polarity or sign. It certainly can vary it need not be constant — but it cannot reverse itself. *a*, *c*, *d*, *e*, *g*, and *h* are all dc. *b* and *f* are known as alternating current, ac.

•18.11

1 is brightest; 2 and 3 are equally bright, though less so than 1; no current passes through 4, 5, or 6, and so they are off altogether.

•18.13

The resistance of the middle branch is half that of the upper branch; it draws twice the current and so 3 is brighter than 1 or 2, which have the same current and are equally dim.

•18.15

R_1 and R_5 have the most current; R_2, R_3 and R_4 have less; R_6 and R_7 have none. (R_7 has no voltage across it and there is no closed circuit through R_6).

Answers to Odd-Numbered Multiple Choice Questions

1. d **3.** d **5.** c **7.** c **9.** a **11.** b **13.** d
15. c **17.** a **19.** b **21.** b

Solutions to Selected Problems

18.1

The voltage (V) across the terminals of a source of emf ($\mathcal{E}$) having an internal resistance (r) is given by Eq. (18.1): $V = \mathcal{E} - Ir$. If the terminals are short-circuited, then there is no voltage difference across them, i.e., $V = \mathcal{E} - Ir = 0$. Plug in $\mathcal{E} = 1.5\,\mathrm{V}$ and $r = 1.0\,\Omega$ and solve for I:

$$I = \frac{\mathcal{E}}{r} = \frac{1.5\,\mathrm{V}}{1.0\,\Omega} = 1.5\,\mathrm{A}\,.$$

18.27

First, use Eq. (18.5) to find the equivalent resistance R_1 of the four resistors ($2\,\Omega$, $3\,\Omega$, $6\,\Omega$, and $1\,\Omega$) in parallel:

$$R_1 = \left(\frac{1}{R_1} + \frac{1}{R_2} + \frac{1}{R_3} + \frac{1}{R_4}\right)^{-1} = \left(\frac{1}{2\,\Omega} + \frac{1}{3\,\Omega} + \frac{1}{6\,\Omega} + \frac{1}{1\,\Omega}\right)^{-1} = 0.5\,\Omega\,.$$

The equivalent resistance between A and B is that of $R_1 = 5\,\Omega$ and $R_2 = 0.5\,\Omega$ in series, so according to Eq. (18.4)

$$R_e = R_1 + R_2 = 5\,\Omega + 0.5\,\Omega = 5.5\,\Omega\,.$$

18.37

The resistance of the lamp can be obtained from the power $\mathrm{P}_{\mathrm{lamp}}$ it consumes under the proper operating voltage V_{lamp}: $\mathrm{P}_{\mathrm{lamp}} = V_{\mathrm{lamp}}^2/R_{\mathrm{lamp}}$, or $R_{\mathrm{lamp}} = V_{\mathrm{lamp}}^2/\mathrm{P}_{\mathrm{lamp}} = (20\,\mathrm{V})^2/80\,\mathrm{W} = 5.0\,\Omega$. Thus if the lamp is placed in series with another resistor R the equivalent resistance of the circuit would be $R_e = R_{\mathrm{lamp}} + R$ which, when connected to a dc source providing an emf $\mathcal{E}$ ($= 60\,\mathrm{V}$), would result in a current of $I = \mathcal{E}/R_e = \mathcal{E}/(R_{\mathrm{lamp}} + R)$. The actual voltage V across the lamp in such a circuit is therefore

$$V = IR_{\mathrm{lamp}} = \frac{R_{\mathrm{lamp}}\mathcal{E}}{R_{\mathrm{lamp}} + R}\,.$$

To operate the lamp properly we need $V = 20\,\mathrm{V}$. Plug this into the formula above, along with $R_{\mathrm{lamp}} = 5.0\,\Omega$ and $\mathcal{E} = 60\,\mathrm{V}$, and solve for R:

$$R = R_{\mathrm{lamp}}\left(\frac{\mathcal{E}}{V_{\mathrm{lamp}}} - 1\right) = (5.0\,\Omega)\left(\frac{60\,\mathrm{V}}{20\,\mathrm{V}} - 1\right) = 10\,\Omega\,.$$

18.41

The current passing through the 17-Ω resistor is $I = 9\,\mathrm{A}$, the same as the reading of the ammeter, since the two are connected in series.

After passing through the 17-Ω resistor I is split into two branches, one passing through $R_1 = 9\,\Omega$ and the other through $R_2 = 1\,\Omega$. Using the result of the previous problem, we have

$$I_1 = I\left(\frac{R_2}{R_1 + R_2}\right) = (9\,\mathrm{A})\left(\frac{1\,\Omega}{9\,\Omega + 1\,\Omega}\right) = 0.9\,\mathrm{A}$$

(the current in the 9-Ω resistor) and $I_2 = I - I_1 = 9\,\mathrm{A} - 0.9\,\mathrm{A} = 8.1\,\mathrm{A} \approx 8\,\mathrm{A}$ (the current in the 1-Ω resistor).

Upon reaching the the next junction, the current has to be split three ways among the three resistors ($R_3 = 3\,\Omega + 1\,\Omega = 4\,\Omega$, $R_4 = 2\,\Omega$, and $R_5 = 4\,\Omega + 1\,\Omega = 5\,\Omega$) in parallel: $I = I_3 + I_4 + I_5$. Also, since they are in parallel the voltage difference across the three resistors must be the same: $I_3 R_3 = I_4 R_4 = I_5 R_5$. To solve these equations for I_3 through I_5, first express I_4 and I_5 in terms of I_3 using the last equation: $I_4 = I_3 R_3 / R_4$, $I_5 = I_3 R_3 / R_5$. Now plug these results into $I = I_3 + I_4 + I_5$ to obtain an equation for I_3: $I = I_3 + I_3 R_3/R_4 + I_3 R_3/R_5 = I_3(1 + R_3/R_4 + R_3/R_5)$, which gives the current in the 3-Ω resistor to be

$$I_3 = \frac{I}{1 + R_3/R_4 + R_3/R_5} = \frac{9\,\mathrm{A}}{1 + 4\,\Omega/2\,\Omega + 4\,\Omega/5\,\Omega} = 2.37\,\mathrm{A} \approx 2\,\mathrm{A}\,.$$

Similarly, the current in the 2-Ω resistor is

$$I_4 = \frac{I}{1 + R_4/R_3 + R_4/R_5} = \frac{9\,\mathrm{A}}{1 + 2\,\Omega/4\,\Omega + 2\,\Omega/5\,\Omega} = 4.74\,\mathrm{A} \approx 5\,\mathrm{A}\,,$$

and that in the 4-Ω resistor is then $I_5 = I - I_3 - I_4 = 9\,\mathrm{A} - 2.37\,\mathrm{A} - 4.74\,\mathrm{A} \approx 2\,\mathrm{A}$.

Alternatively, the values of I_3 through I_5 can also be obtained by first finding the equivalent resistance of the three-resistor combination, which is $\frac{20}{19}\,\Omega$, then finding the voltage difference across each of the resistor, at $V = \left(\frac{20}{19}\,\Omega\right)(9\,\mathrm{A}) = 9.474\,\mathrm{V}$. The current in each branch then follows directly from $I_3 = V/R_3$, etc.

Finally, all the current (9 A) flows through the 23-Ω resistor which is in series with the ammeter.

18.59

According to the discussion on *R-C* circuit in the text the time it takes for a capacitor C connected in series with a resistor R to charge to 63% of its maximum range is equal to RC, the time constant of the circuit. In our case $R = 5.0\,\mathrm{k\Omega} = 5.0 \times 10^3\,\Omega$ and $C = 800\,\mu\mathrm{F} = 800 \times 10^{-6}\,\mathrm{F}$; so the charging time in question is

$$RC = (5.0 \times 10^3\,\Omega)(800 \times 10^{-6}\,\mathrm{F}) = 4.0\,\mathrm{s}\,.$$

18.75

Points A and C are connected directly with the two terminals of the 12-V battery, and there is no more power source in the *A-D-C* segment of the circuit; so the voltage across *A-D-C* is just 12 V. With a resistance of the $2\,\Omega + 2\,\Omega = 4\,\Omega$, the current in *A-D-C* is then $12\,\mathrm{V}/4\,\Omega = 3\,\mathrm{A}$, which flows from A through D to C.

Now consider the segment A-B-C. Due to the 6-V battery present the net voltage in the loop A-B-C-A is $12\,\text{V} - 6\,\text{V} = 6\,\text{V}$. Divide this by $2\,\Omega$, the resistance in A-B-C, to obtain the current in the segment to be $I = 6\,\text{V}/2\,\Omega = 3\,\text{A}$, which flows from A through B to C.

Finally, apply the Node Rule to point A. We learned from the calculation above that two branches of current flow out of point A, at 3 A each. The current flowing into point A from the 12-V battery must therefore be $3\,\text{A} + 3\,\text{A} = 6\,\text{A}$. This is the current in A-C.

18.87

Apply Kirchhoff's Loop Rule to the right half of the circuit (including the middle and right vertical branch), going counterclockwise from the the positive terminal of the 48-V battery:

$$-(3.0\,\text{A})(4.0\,\Omega) - (1.0\,\text{A})R_1 + 10\,\text{V} + 48\,\text{V} = 0\,,$$

which gives $R_1 = 46\,\Omega$. Now apply the Loop Rule again, this time around the perimeter of the circuit (including both the left and the right vertical branches), going counterclockwise from the the positive terminal of the 48-V battery:

$$-(3.0\,\text{A})(4.0\,\Omega) - (2.0\,\text{A})(25\,\Omega) + \mathcal{E}_1 + 48\,\text{V} = 0\,,$$

which gives $\mathcal{E}_1 = 14\,\text{V}$.

18.93

It is clear from the Node Rule that $I_6 = I_2 = 2.0\,\text{A}$ and $I_5 = I_4 = 5.0\,\text{A}$. Now apply the same rule to the node at the top of the circuit where I_1 splits into three branches:

$$I_1 = I_2 + I_3 + I_4 = 2.0\,\text{A} + 1.0\,\text{A} + 5.0\,\text{A} = 8.0\,\text{A}\,.$$

To find V, apply the Loop Rule to the part of the circuit containing I_1 and I_3, going clockwise from the positive terminal of the battery:

$$-I_3(30\,\Omega) + V = 0\,,$$

or $V = I_3(30\,\Omega) = (1.0\,\text{A})(30\,\Omega) = 30\,\text{V}$.

To find R, do the same for the loop containing I_1, I_4 and I_5:

$$-I_4(2.0\,\Omega) - I_5R + V = 0\,.$$

Plug in $I_4 = I_5 = 5.0\,\text{A}$ and $V = 30\,\text{V}$ to find $R = 4.0\,\Omega$.

19 *Magnetism*

Answers to Selected Discussion Questions

•19.1

The electrostatic E-field of a point charge should have spherical symmetry because the charge presumably has spherical symmetry and space is isotropic. Once the charge is set moving, the axis of that motion represents a distinguishable direction and both the E and B fields will now only be axially symmetric (one circular, one radial).

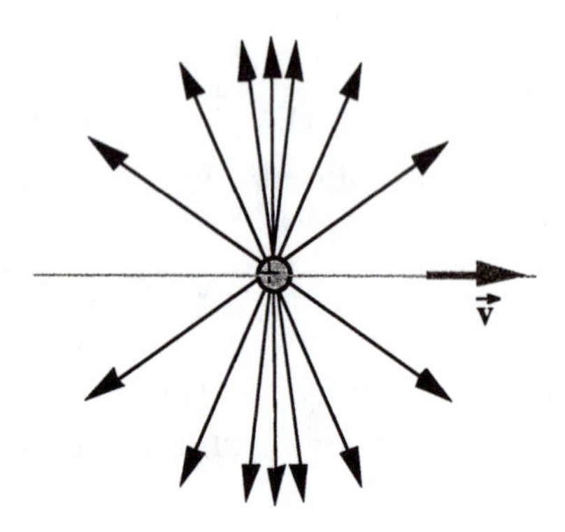

•19.3

Each north pole would experience a torque of $q_m B_N R_N$ and each south pole a torque of $q_m B_S R_S$ in the opposite direction. These must cancel since there's no rotation, hence they are equal and $B_S/B_N = R_N/R_S$; the field varies inversely with the radial distance.

•19.7

The field of the coil rapidly changes magnitude and direction and that tends to mix up the domain structure. Since $B \rightarrow 0$ as the coil is removed, the domains cannot continue to realign themselves with the field of the coil.

•19.11

Far from any long wiggly wire the field must resemble that of a straight wire. Certainly if the bends are perfectly regular (e.g., an equilateral sawtooth pattern or a sinusoid) we could expect the same field (some distance away) as from a straight current. Since field lines do not cross we can imagine sets of circular lines around each tiny segment of the bent wire. These can be thought of as flattening out into planes perpendicular to the axis of the wire at distances large compared to the wiggles remember the E-field of two equal charges. Ampère's Law makes no distinction between wiggly currents and straight ones. Experimentally, it is found that the setup will be unaffected by the introduction of a B-field. A solenoid is a spiralling wire, and if it has a projected length (end-to-end), it will have the external field of a straight wire.

•19.15

Looking down on the left side, there is a counter-clockwise B-field and the north pole rotates counterclockwise around with the field. On the right, there is a downward current in the B-field of the magnet, and therefore the hanging rod experiences a perpendicular force that causes it to rotate clockwise.

•19.17

The magnet's field was expelled from the disk when the latter went superconducting. The result is very much as if the field lines were bent up and away by the presence of an identical magnet (with like poles) within the superconductor that repels the cube.

Answers to Multiple Odd-Numbered Choice Questions

1. d **3.** c **5.** c **7.** b **9.** b **11.** a **13.** a
15. b **17.** c **19.** c **21.** a

Solutions to Problems

19.13

The B-field inside a long solenoid with n turns per unit length carrying a current I is given by Eq. (19.5): $B_z \approx \mu_0 nI$. In our case the solenoid is 10 cm (0.10 m) long with a total of

200 turns, so $n = 200/0.10\,\mathrm{m} = 2.0 \times 10^3\,\mathrm{m}^{-1}$. Plug this into the formula for B_z above, along with $B_z = 0.50\,\mathrm{mT} = 5.0 \times 10^{-4}\,\mathrm{T}$, and solve for the current I needed:

$$I \approx \frac{B_z}{\mu_0 n} = \frac{5.0 \times 10^{-4}\,\mathrm{T}}{(4\pi \times 10^{-7}\,\mathrm{T{\cdot}m/A})(2.0 \times 10^3\,\mathrm{m}^{-1})} = 0.20\,\mathrm{A}\,.$$

19.19

Each electron carries a charge of $q_e = 1.602 \times 10^{-19}\,\mathrm{C}$ (absolute value). If the number of electrons arriving at the screen per unit time is n_e, then the total amount of charge delivered onto the screen by the electron beam per unit time is $n_e q_e$ which, by definition, is the current carried I by the beam. treating such a current as long and straight (which is a valid assumption here), we may use Eq. (19.2) to find the magnetic field B a distance r from the beam: $B = \mu_0 I/2\pi r = \mu_0(n_e q_e)/2\pi r$. Plug in $n_e = 6.0 \times 10^{12}\,\mathrm{s}^{-1}$, $q_e = 1.602 \times 10^{-19}\,\mathrm{C}$, and $r = 1.5\,\mathrm{cm} = 1.5 \times 10^{-2}\,\mathrm{m}$ to obtain

$$\begin{aligned} B &= \frac{\mu_0 n_e q_e}{2\pi r} \\ &= \frac{(4\pi \times 10^{-7}\,\mathrm{T{\cdot}m/A})(6.0 \times 10^{12}\,\mathrm{s}^{-1})(1.602 \times 10^{-19}\,\mathrm{C})}{2\pi(1.5 \times 10^{-2}\,\mathrm{m})} \\ &= 1.3 \times 10^{-11}\,\mathrm{T}\,. \end{aligned}$$

Since the electrons carry negative charge the direction of the current is *away* from the screen. You can see that the B-field is oriented clockwise looking towards the source from the screen (i.e., facing the beam of the electrons).

19.27

At the center of the circular loop $z = 0$, and the expression for B_z given in the problem statement reduces to

$$B_z = \left.\frac{\mu_0 I R^2}{2\,(R^2 + z^2)^{3/2}}\right|_{z=0} = \frac{\mu_0 I R^2}{2\,(R^2)^{3/2}} = \frac{\mu_0 I}{2R}\,,$$

which coincides with Eq. (19.3). At $z \gg R$ the R^2 term in the denominator in the expression for B_z becomes negligible compared with z^2, so the denominator is now $2(R^2 + z^2)^{3/2} \approx 2(z^2)^{3/2} = 2z^3$, whereupon

$$B_z = \frac{\mu_0 I R^2}{2\,(R^2 + z^2)^{3/2}} \approx \frac{\mu_0 I R^2}{2z^3} \qquad \text{for } z \gg R\,.$$

19.43

Consider an Ampèrian loop in the shape of a circle of radius r, in the same plane, and concentric with, the toroid. By symmetry the B-field anywhere along the loop has the same magnitude and is tangential in direction (i.e., perpendicular to the radial direction) — if it exists at all. Thus Ampère's Law reads

$$\oint B_{\parallel}\,dl = \oint B\,dl = B\oint dl = B(2\pi r) = \mu_0 \sum I\,,$$

which yields

$$B = \frac{\mu_0 \sum I}{2\pi r}\,.$$

Note that here $\sum I$ is the current enclosed by the Ampèrian loop. For $r < R_i$ it is clear that $\sum I = 0$, since no current is encompassed by the loop. Similarly $\sum I = 0$ for $r > R_o$ as well, since the number of wires carrying current into the circular area enclosed by the Ampèrian loop is the same as that out of the loop. Hence

$$B \propto \sum I = 0$$

except for the region inside the confines of the toroid (where $R_i < r < R_o$).

19.53

First of all, the only two segments which contribute to the B-field at point P are the straight one with length L and the curved arc; no other segments contribute since θ, the angle between $d\vec{\mathbf{l}}$ and $\hat{\mathbf{r}}$, is zero [and therefore according to Eq. (19.8) $dB \propto \sin\theta = 0$]. To find B_1, the contribution from the straight wire of length L, refer to Example (19-3), in which the B-field of an *infinitely* long wire is found to be

$$B = \frac{\mu_0 I}{4\pi x}\int_0^{\pi} \sin\theta\,d\theta$$

at a point P a perpendicular distance x from the wire. In the present case the only differences are that x is replaced by h, and that the length of the wire is finite, with the left end of the wire making an angle of $\pi/2$ with $\vec{\mathbf{r}}$ while the right end at an angle $\pi - (\pi/2 - \pi/6) = 2\pi/3$ with $\vec{\mathbf{r}}$. Thus the integral over θ should be from $\pi/2$ to $2\pi/3$ instead of from 0 to π:

$$B_1 = \frac{\mu_0 I}{4\pi h}\int_{\pi/2}^{2\pi/3} \sin\theta\,d\theta = \frac{\mu_0 I}{4\pi h}\Big[-\cos\theta\Big]_{\pi/2}^{2\pi/3} = \frac{\mu_0 I}{4\pi h}\frac{\sqrt{3}}{2} = \frac{\mu_0 I}{8\pi L}\,,$$

where in the last step we noted that $h = \sqrt{3}L$. $\vec{\mathbf{B}}_1$ points out of the page.
For the B-field B_2 of the curved arc $dB_2 = (\mu_0/4\pi)(I d\varphi/R)$, where R is the radius of the circle which contains the arc in question. Integrating over φ from 0 to $\pi/6$ to obtain

$$B_2 = \int_0^{\pi/6} \frac{\mu_0}{4\pi}\frac{I d\varphi}{R} = \frac{\mu_0}{4\pi}\frac{I}{R}\int_0^{\pi/6} d\varphi = \frac{\mu_0}{4\pi}\frac{I(\pi/6)}{R} = \frac{\mu_0 I}{24R}\,,$$

into the page. The net B-field at point P is then

$$B = B_s - B_c = \frac{\mu_0 I}{8\pi L} - \frac{\mu_0 I}{24R}\,,$$

out of the page.

19.57

Label the top and bottom sheet with subscripts 1 and 2, respectively. From the result of Problem (19.45) we know that the magnitude of the B-field due to either plate is $B_1 = B_2 = \frac{1}{2}\mu_0 i$, and $\vec{\mathbf{B}}_1$ points due south above sheet 1 and north below it while $\vec{\mathbf{B}}_2$ points due north above sheet 2 and south below it. Thus in the region above both sheets

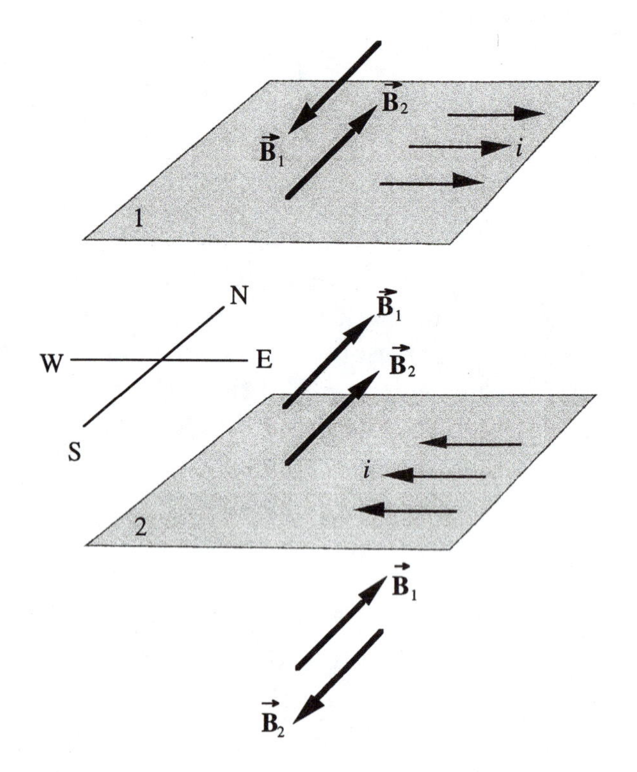

$$\begin{aligned}\vec{\mathbf{B}} &= \vec{\mathbf{B}}_1 + \vec{\mathbf{B}}_2 \\ &= \left(\frac{1}{2}\mu_0 i\right)\text{-south} + \left(\frac{1}{2}\mu_0 i\right)\text{-north} \\ &= 0,\end{aligned}$$

in the region in between the two sheets (i.e., below sheet 1 and above sheet 2)

$$\vec{\mathbf{B}} = \vec{\mathbf{B}}_1 + \vec{\mathbf{B}}_2 = \left(\frac{1}{2}\mu_0 i\right)\text{-north} + \left(\frac{1}{2}\mu_0 i\right)\text{-north} = (\mu_0 i)\text{-north},$$

and in the region below both sheets

$$\vec{\mathbf{B}} = \vec{\mathbf{B}}_1 + \vec{\mathbf{B}}_2 = \left(\frac{1}{2}\mu_0 i\right)\text{-north} + \left(\frac{1}{2}\mu_0 i\right)\text{-south} = 0.$$

19.63

Start with $F_M = q_\bullet vB\sin\theta$ [Eq. (19.10)], where in the SI system F_M is in N, $q_\bullet$ in C, v in m/s, and B in T. Equate the SI units of both sides of the equation above:

$$\mathrm{N} = \mathrm{C}\cdot(\mathrm{m/s})\cdot\mathrm{T} = (\mathrm{C/s})\cdot(\mathrm{m}\cdot\mathrm{T}) = \mathrm{A}\cdot(\mathrm{m}\cdot\mathrm{T})$$

(note that $1\,\mathrm{A} = 1\,\mathrm{C/s}$). Equate the first and the last entries in the equation above: $\mathrm{N} = \mathrm{A}\cdot(\mathrm{m}\cdot\mathrm{T})$, and divide both sides by A^2 to obtain

$$\mathrm{N/A}^2 = \frac{\mathrm{A}\cdot(\mathrm{m}\cdot\mathrm{T})}{\mathrm{A}^2} = \frac{\mathrm{m}\cdot\mathrm{T}}{\mathrm{A}} = \mathrm{T}\cdot\mathrm{m/A}.$$

19.87

The configuration described in the problem statement is depicted to the right. The current in the wire can be effectively thought of as the result of a collection of positively charged particles moving with a velocity $\vec{\mathbf{v}}$ in the direction of the current flow. Consider one such charge (q). The magnetic force $\vec{\mathbf{F}}_M$ exerted on the charge is given by Eq. (19.10) to be $\vec{\mathbf{F}}_M = q\vec{\mathbf{v}} \times \vec{\mathbf{B}}$, which is in the negative y-direction (see the figure to the right). This is also the direction of the total magnetic force on the wire.

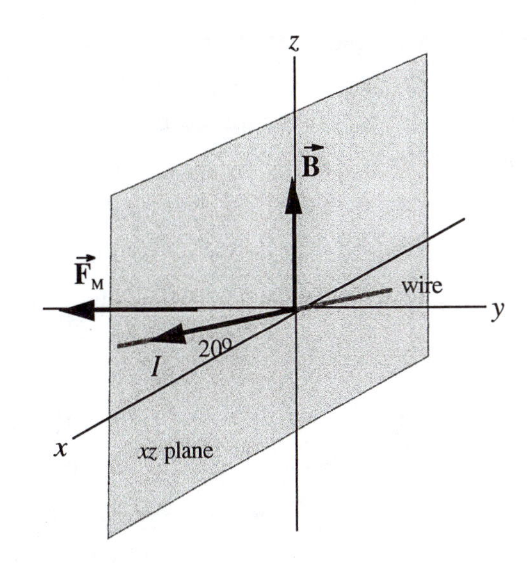

19.101

Since the wire is long we may use Eq. (19.17), which gives the magnetic force per unit length between two parallel, charge-carrying wires: $F_M/l = \mu_0 I_1 I_2 / 2\pi d$. In this case $I_1 = I_2 = 10\,\mathrm{A}$, while d, the separation between the two parallel segments of the wire, is equal to the diameter of the tree, at $2 \times 1.0\,\mathrm{m} = 2.0\,\mathrm{m}$. Thus

$$\frac{F_M}{l} = \frac{\mu_0 I_1 I_2}{2\pi d} = \frac{(4\pi \times 10^{-7}\,\mathrm{T\cdot m/A})(10\,\mathrm{A})(10\,\mathrm{A})}{2\pi(2.0\,\mathrm{m})} = 2.0 \times 10^{-5}\,\mathrm{N/m}\,.$$

The force is repulsive, since the current flows from the battery towards the tree in one segment of the wire and backwards (from the tree back to the battery) in the other.

19.103

Refer to the diagram to the right. Consider an infinitesimal current element $Id\vec{\mathbf{l}}$ of the rod, located between x and $x + dx$. The magnetic field $\vec{\mathbf{B}}$ due to the bar magnet at this location points directly from the north pole of the bar to the current element in question, with a magnitude of $B = C/r^2$, where by the Pythagorean Theorem $r^2 = h^2 + x^2$. The magnetic force $d\vec{\mathbf{F}}_M$ exerted by the magnet on the current element is then $d\vec{\mathbf{F}}_M = Id\vec{\mathbf{l}} \times \vec{\mathbf{B}}$, pointing perpendicularly out of the page (assuming that the current in the rod flows from left to right). The magnitude of the magnetic force is

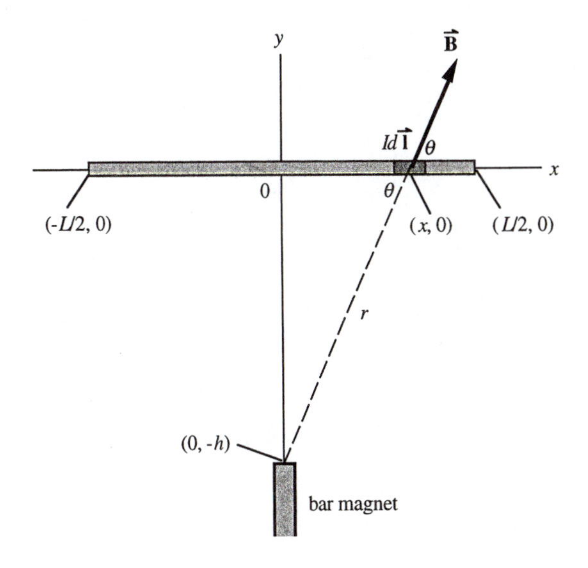

$$dF_M = |Id\vec{\mathbf{l}} \times \vec{\mathbf{B}}| = (Idx)B\sin\theta = (Idx)\left(\frac{C}{r^2}\right)\left(\frac{h}{r}\right) = \frac{ICh\,dx}{(h^2 + x^2)^{3/2}}\,,$$

where we noted that $dl = dx$, $B = C/r^2$, $\sin\theta = h/r$, and $r = \sqrt{h^2 + x^2}$. Since the direction of $d\vec{\mathbf{F}}_{\text{M}}$ is the same throughout the rod we may just integrate over its magnitude dF_{M} to obtain that of the magnetic force on the entire rod. Noting that x varies from $-L/2$ to $+L/2$ for the rod, we have

$$F_{\text{M}} = \int dF_{\text{M}} = \int_{-L/2}^{L/2} \frac{ICh}{(h^2 + x^2)^{3/2}}\, dx\,.$$

Upon making a change of variable $x = h\tan\phi$ and using the identities $1 + \tan^2\phi = \sec^2\phi = 1/\cos^2\phi$ and $d\tan\phi = \sec^2\phi\, d\phi$, we find the integral

$$\begin{aligned}\int \frac{dx}{(h^2 + x^2)^{3/2}} &= \int \frac{d(h\tan\phi)}{(h^2 + h^2\tan^2\phi)^{3/2}} = \frac{1}{h^2}\int \frac{\sec^2\phi\, d\phi}{\sec^3\phi} = \frac{1}{h^2}\int \cos\phi\, d\phi \\ &= \frac{\sin\phi}{h^2} + \text{const.} = \frac{x}{h^2\sqrt{h^2 + x^2}} + \text{const.}\end{aligned}$$

Thus

$$F_{\text{M}} = ICh\left[\frac{x}{h^2\sqrt{h^2 + x^2}}\right]_{-L/2}^{L/2} = \frac{ICL}{h\sqrt{h^2 + (L/2)^2}}\,.$$

20 *Electromagnetic Induction*

Answers to Selected Discussion Questions

•20.3

The handle of the device slips into a coil hidden in the base of the holder. Another coil is inside the handle and the two are in close proximity. Connecting the coil in the well to AC generates a time-varying B-field that passes through the plastic skin of the handle and produces a rapidly varying (60 Hz) induced emf and an induced current. That current is converted to DC and used to charge a battery in the handle, which powers the motor in the device. No charge and so no "electricity" passes from the base to the handle, although electrical energy certainly is transferred. The same scheme will work across human skin to power a mechanical heart.

•20.5

As the magnet approaches, its B-field attempts to penetrate the ring. A supercurrent is thereby induced opposing the buildup of flux and the motion of the magnet. If the ring is free to move, it will lift off its support and hover (with its induced north pole down) above the north pole of the magnet. Finite, persistent supercurrents on the surface of the superconductor circulate in such a way as to shield the interior from the field. No flux enters the body of the superconductor; there is no change in flux, no E-field induced, and no bulk current.

•20.7

To crank the generator without a load and therefore without a current (in the steady state), one need only overcome friction. By contrast, lighting the 100-W bulb requires twice the power used in lighting the 50-W bulb and that power must be supplied by the person turning the crank. Most people will find it difficult to keep the 100-W bulb lit very long.

•20.9

Current would be induced in the loop and power transferred from the line to it — the time-varying B-field would induce an E-field. If a wire loop was in that region, the E-field would do work on the free electrons, transferring energy to them. The power company could detect a loss in energy arriving at the end of the line; there would be a slight drop in delivered current since the voltage is fixed by the generator. If you separate the two leads from a telephone and lay the pick-up loop (attached to a good set of headphones) next to one of the wires, you should be able to listen in on the time-varying B-field of the conversation.

•20.13

Yes, a coil in a motor turning through a magnetic field will experience an induced emf and an induced current that will send energy back to the source, which is driving the motor. With no load, the motor produces (and returns to the power company) almost as much current as it draws, and so costs very little to run. A free-turning motor must have its speed limited by the back-current it generates with no losses (in the bearings, etc.) the motor will speed up until the back-current equals the driving current and it can no longer accelerate. With a load, the motor does work and draws energy in excess of what it returns via the back-current. In real life, a motor also produces a good deal of thermal energy via friction and if it is to operate continuously for any long period of time, it must be cooled (usually by forced air). A jammed motor will not turn and not generate a back-current. The driving current (which depends on the motor's resistance) now undiminished by any back-current is too great. The wiring will heat up and the insulation will begin to burn off. If the process is not stopped soon, the motor will be destroyed.

•20.15

Recall that the B-field outside a very long tight solenoid approaches zero. Each meter reads the potential drop across the resistor adjacent to it, the resistor with which it forms a closed circuit excluding the solenoid. The induced emf equals the difference between the two meter readings, or alternatively, the sum of the potential differences across the resistors in the central circuit (1-9-10-2-7-8-1). The seemingly strange thing here is that the "voltage" measured between 1 and 2 actually depends on how you hook up the meter. The varying flux provides energy to the induced current (energy coming from the power source driving current through the solenoid). In part (b) both meters would read $I_i R_2$.

Answers to Odd-Numbered Multiple Choice Questions

1. e **3.** c **5.** a **7.** a **9.** c **11.** c **13.** b
15. d **17.** d **19.** b **21.** b

Solutions to Selected Problems

20.3

The flux density B is, by definition, the magnetic flux Φ_M per unit cross-sectional area: $B = \Phi_M/A$, where A is the total cross-sectional area perpendicular to the B-field [see Eq. (20.1)]. Plug in $\Phi_M = 6.0\,\text{mWb} = 6.0 \times 10^{-3}\,\text{Wb}$ and $A = 50\,\text{cm2} = 50 \times 10{-}^{4}\,\text{m}^2$ to obtain

$$B = \frac{\Phi_M}{A} = \frac{6.0 \times 10^{-3}\,\text{Wb}}{50 \times 10^{-4}\,\text{m}^2} = 1.2\,\text{Wb/m}^2 = 1.2\,\text{T}\,.$$

20.7

The change in the magnetic field is

$$\Delta\vec{\mathbf{B}} = \vec{\mathbf{B}}_f - \vec{\mathbf{B}}_i = 0 - (0.01\,\text{T})\text{-east} = -(0.01\,\text{T})\text{-east} = (0.01\,\text{T})\text{-west}\,,$$

i.e., $\Delta\vec{\mathbf{B}}$ has a magnitude of 0.01 T and is due west.

20.9

Before the loop is yanked from the B-field the magnetic flux through the loop is $\Phi_{Mi} = BA_\perp$, where $A_\perp$ is the encompassed area of the loop perpendicular to the B-field. In a time interval Δt the B-field through the loop is gone so $\Phi_{Mf} = 0$. The induced emf, which is equal in magnitude to the rate of change of the magnetic flux in the loop, is then [see Eq. (20.3)]

$$\mathcal{E} = -N\frac{\Delta\Phi_M}{\Delta t} = -N\left(\frac{\Phi_{Mf} - \Phi_{Mi}}{\Delta t}\right) = -N\left(\frac{0 - BA_\perp}{\Delta t}\right) = \frac{NBA_\perp}{\Delta t}\,.$$

Plug in $N = 1$, $B = 0.40\,\text{T}$, $A_\perp = 0.25\,\text{m}^2$, and $\Delta t = 200\,\text{ms} = 0.200\,\text{s}$ to obtain $\mathcal{E} = +0.50\,\text{V}$. (Here the plus sign in $\mathcal{E}$ means that the induced emf would generate a current which in turn produces a B-field whose flux is positive through the loop, as is Φ_{Mi}.)

20.23

First, compute the induced emf in the coil from Eq. (20.3): $\mathcal{E} = -Nd\Phi_M/dt$, where according to Eq. (20.1) $\Phi_M = BA_\perp$. In this case $A_\perp$ is a constant while B changes at the rate of dB/dt, so $d\Phi_M/dt = d(BA_\perp)/dt = A_\perp(dB/dt)$; and so the magnitude of the induced emf is $|\mathcal{E}| = NA_\perp|dB/dt|$.

When a steady state is reached in the coil there is no more charging or discharging going on across the capacitor so $|\mathcal{E}|$ must be balanced by the voltage difference V_C across the capacitor: $|\mathcal{E}| = V_C = Q/C$, where Q is the steady-state charge on the capacitor (of capacitance C). Combine this equality with the expression for $|\mathcal{E}|$ obtained above to yield

$$|\mathcal{E}| = NA_{\perp}\left|\frac{dB}{dt}\right| = V_C = \frac{Q}{C}.$$

Substitute $N = 220$, $|dB/dt| = 20\,\text{mT/s}$, $A_{\perp} = \pi(10\,\text{cm})^2$, and $C = 30\,\mu\text{F}$ into this equation and solve for Q:

$$Q = CNA_{\perp}\left|\frac{dB}{dt}\right| = (30\,\mu\text{F})(220)\left[\pi(0.10\,\text{m})^2\right](20\times10^{-3}\,\text{T/s}) = 4.1\,\mu\text{C}.$$

20.27

In Problem (20.24) we proved the formula for ΔQ, the charge that flows through a galvanometer in response to a magnetic flux change: $\Delta Q = N|\Delta\Phi_M|/R$. We now find the expression for $\Delta\Phi_M$ in this case. The initial magnetic flux due to the B-field inside the solenoid through the coil of area A before the current is reversed is $\Phi_{Mi} = BA$, while afterwards it changes to $\Phi_{Mf} = -BA$. (Here we have arbitrarily chosen the initial value of the flux to be positive so the final value is negative.) The magnitude of the change in Φ_M is then $|\Delta\Phi_M| = |\Phi_{Mf} - \Phi_{Mi}| = |-BA - BA| = 2BA$. Plug this expression for $|\Delta\Phi_M|$ into the formula for ΔQ to yield

$$\Delta Q = \frac{N|\Delta\Phi_M|}{R} = \frac{N(2BA)}{R},$$

which we solve for B:

$$B = \frac{R\Delta Q}{2NA} = \frac{(0.50\,\Omega)(2.0\times10^{-6}\,\text{C})}{2(12)(0.50\times10^{-4}\,\text{m}^2)} = 8.3\times10^{-4}\,\text{T} = 0.83\,\text{mT}.$$

20.37

The bulb would not light, nor would a voltmeter connected across the wing tips be able to pick up a reading indicating the voltage difference calculated in Problem (20.33). In fact the leads of the meter also cut through the Earth's B-field at the same rate as the wings so the same amount of motional emf would appear in the voltmeter as well as the leads; and once they are attached to the wings the closed loop consisting of the wings, the leads and the voltmeter would have two oppositely directed emf of the same strength, so no current can circulate in the circuit.

Alternatively, note that the flux through the closed loop of meter-leads-wings does not change if the B-field remains constant so no net emf appears in the circuit to drive a current: $\mathcal{E} \propto \Delta\Phi_M = 0$. You could, however, read a voltage difference if the meter stayed on the ground and still managed to remain connected to the plane.

20.49

The diagram to the right is a top view of the rod as it rotates at an angular speed ω about point O, with the magnetic field perpendicular to the page, in which the rod rotates. During a time interval t the rod sweeps out an angular displacement $\theta = \omega t$, and the area of the imaginary, fan-shaped loop depicted to the right is $A = (\theta/2\pi)(\pi L^2) = \theta L^2/2$, which is clear since the area is a fraction $\theta/2\pi$ of the circular area πL^2 that the rod would sweep out in one complete revolution. The magnetic flux through the loop is then

$$\Phi_{\mathrm{M}} = BA = B\left(\frac{1}{2}\theta L^2\right) = \frac{1}{2}BL^2\omega t\,.$$

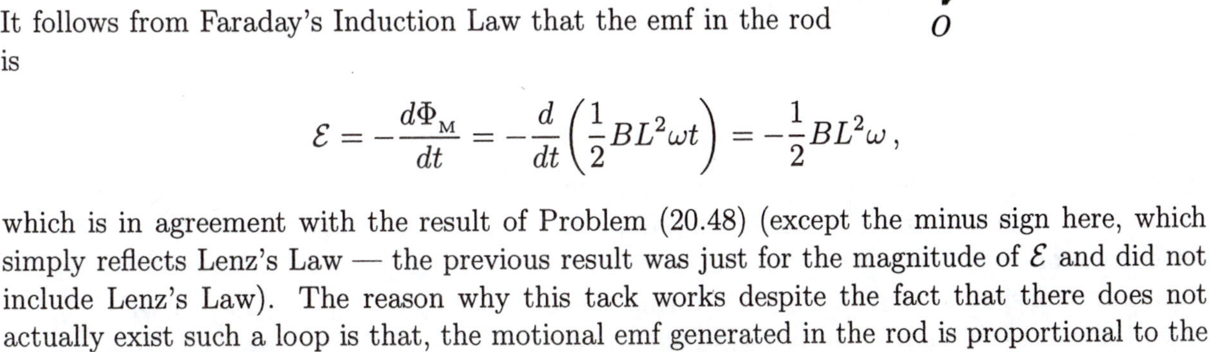

It follows from Faraday's Induction Law that the emf in the rod is

$$\mathcal{E} = -\frac{d\Phi_{\mathrm{M}}}{dt} = -\frac{d}{dt}\left(\frac{1}{2}BL^2\omega t\right) = -\frac{1}{2}BL^2\omega\,,$$

which is in agreement with the result of Problem (20.48) (except the minus sign here, which simply reflects Lenz's Law — the previous result was just for the magnitude of $\mathcal{E}$ and did not include Lenz's Law). The reason why this tack works despite the fact that there does not actually exist such a loop is that, the motional emf generated in the rod is proportional to the rate at which magnetic field lines are bing cut, which in turn is equal to the rate at which the magnetic flux through the imaginary loop is changing.

20.59

The standard form of the emf produced by an ac generator as a function of times is given by Eq. (20.7): $\mathcal{E} = NAB\omega\sin\omega t$, where N is the number of turns of the coil, A is its area, B is the magnetic field, and $\omega = 2\pi f$ is the angular speed of rotation of the coil in the B-field. Here f is the frequency of the output emf. Note that the maximum emf $\mathcal{E}_{\mathrm{m}}$ occurs when $|\sin\omega t|$ reaches its peak value of 1, whereupon $\mathcal{E} = \mathcal{E}_{\mathrm{m}} = NAB\omega$. Thus we may also write $\mathcal{E} = \mathcal{E}_{\mathrm{m}}\sin\omega t$. Comparing this standard form with the expression given: $\mathcal{E} = 100\sin 376.99t$, we get, in this case, $\mathcal{E}_{\mathrm{m}} = 100\,\mathrm{V}$ and $\omega t = 2\pi f t = 376.99t$, which yields

$$f = \frac{376.99\,\mathrm{Hz}}{2\pi} = 60.000\,\mathrm{Hz}\,.$$

20.61

The angle between the direction of the magnetic field $\vec{\mathbf{B}}$ and the normal to the plane of the loop is initially zero. At time t into the rotation the angle becomes $\theta = \omega t$, as the plane of the loop has been in rotation at an angular speed ω for a time t. The magnetic flux through the loop at that moment is

$$\Phi_{\text{M}} = \int \vec{\mathbf{B}} \cdot d\vec{\mathbf{A}} = \int B\cos\theta \, dA = B\cos\theta \int dA = AB\cos\theta = (\pi R^2)B\cos\omega t\,,$$

where we noted that B is uniform (so it can be taken out of the integration), and that the surface area of a circular loop of radius R is $A = \pi R^2$. According to Faraday's Induction Law [Eq. (20.2)] the induced emf in the loop is then

$$\begin{aligned}\mathcal{E} &= -\frac{d\Phi_{\text{M}}}{dt} = -\frac{d}{dt}\left(\pi R^2 B\cos\omega t\right) \\ &= -\pi R^2 B\left(\frac{d\cos\omega t}{dt}\right) = -\pi B R^2\left(-\omega\sin\omega t\right) \\ &= \pi R^2 \omega B\sin\omega t\,.\end{aligned}$$

20.63

Following the hint given in the problem statement, we first find $\Delta A/\Delta t$, the rate at which a radial strip of conductor sweeps out an area as a result of its rotation. Consider a time interval Δt, during which the disk rotates through an angle $\Delta\theta$, which satisfies $\Delta\theta = \omega\Delta t$, with ω the angular speed of rotation for the disk. During that time interval the leads connected to the resistor between the axis and the rim also sweeps through the same angle, covering an area ΔA as shown to the right. Note that ΔA is a fraction of the total area A of the disk of radius r: $\Delta A/A = \Delta\theta/2\pi$, and so

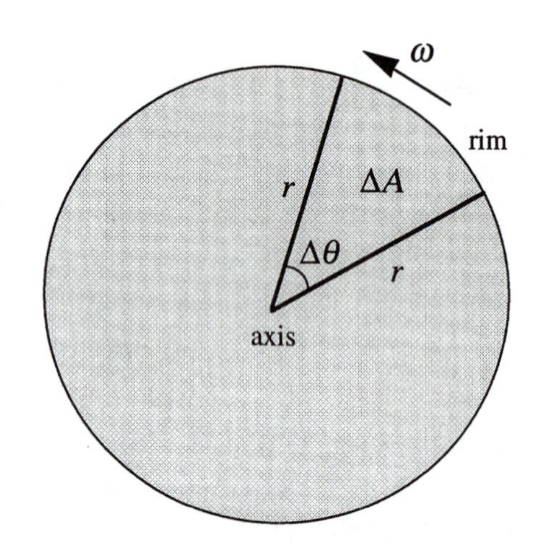

$$\Delta A = A\left(\frac{\Delta\theta}{2\pi}\right) = \pi r^2\left(\frac{\Delta\theta}{2\pi}\right) = \frac{1}{2}r^2\Delta\theta = \frac{1}{2}r^2\omega\Delta t\,.$$

The rate of change of magnetic flux due to the rotation of the resistor is then

$$\mathcal{E} = \frac{\Delta\Phi_{\text{M}}}{\Delta t} = \frac{B\Delta A}{\Delta t} = \frac{r^2 B\omega\Delta t}{2\Delta t} = \frac{1}{2}r^2\omega B\,,$$

which is equal in magnitude to the induced emf $\mathcal{E}$ across the resistor. The induced current I in the resistor follows as $I = \mathcal{E}/R = \frac{1}{2}r^2\omega B/R$. Substitute $I = 1.25\,\text{mA}$, $r = 25\,\text{cm}$, $\omega = 360\,\text{rpm} = (360 \times 2\pi)/60\,\text{s} = 37.7\,\text{rad/s}$, and $R = 20\,\Omega$ into this equation and solve for B:

$$B = \frac{2IR}{\omega r^2} = \frac{2(1.25\times10^{-3}\,\text{A})(20\,\Omega)}{(37.7\,\text{rad/s})(0.25\,\text{m})^2} = 2.1\times10^{-2}\,\text{T} = 21\,\text{mT}\,.$$

20.101

(a) In an R-L circuit the voltage output V of the power source is related to the current I in the circuit via Eq. (20.11): $V = L(dI/dt) + IR$. In the steady state there is no change in current, i.e., $dI/dt = 0$, and so the steady-state current I_{m} in the circuit satisfies $V = I_{\mathrm{m}}R$, or

$$I_{\mathrm{m}} = \frac{V}{R} = \frac{12\,\mathrm{V}}{6.0\,\Omega} = 2.0\,\mathrm{A}\,.$$

(b) The time constant is given by

$$\frac{L}{R} = \frac{24\,\mathrm{H}}{6.0\,\Omega} = 4.0\,\mathrm{s}\,.$$

(c) The current in the R-L circuit varies as a function of time as $I = I_{\mathrm{m}}\left(1 - e^{-tR/L}\right)$, which we rewrite as $e^{-tR/L} = 1 - I/I_{\mathrm{m}}$. To solve for the time t for I to reach within 1% of I_{m} (i.e., when $I/I_{\mathrm{m}} = 1 - 1\% = 0.99$), take the natural logarithm of both sides to obtain $\ln\left(e^{-tR/L}\right) = -tR/L = \ln(1 - I/I_{\mathrm{m}})$, or

$$t = -\frac{L}{R}\ln\left(1 - \frac{I}{I_{\mathrm{m}}}\right) = -(4.0\,\mathrm{s})\ln(1 - 0.99) = 18\,\mathrm{s} \approx 0.2 \times 10^2\,\mathrm{s}\,.$$

Note that we made use of the identity $\ln e^x = x$.

20.113

The magnetic field B inside a long solenoid of cross-sectional area A with n turns per unit length is $B = \mu_0 nI$ as a current I flows through it. The magnetic energy density inside the solenoid then follows from Eq. (20.16) to be $u_{\mathrm{M}} = \frac{1}{2}B^2/\mu_0 = \frac{1}{2}(\mu_0 nI)^2/\mu_0 = \frac{1}{2}\mu_0 n^2 I^2$. Now consider a segment of the solenoid with a cross-sectional area A and length l. The volume of the segment is $V = Al$, and the total magnetic energy stored in the segment is

$$\mathrm{PE_M} = u_{\mathrm{M}}V = \left(\frac{1}{2}\mu_0 n^2 I^2\right)Al\,.$$

Plug in $n = 2.0/\mathrm{mm} = 2.0 \times 10^3/\mathrm{m}$, $I = 5.0\,\mathrm{A}$, and $A = \pi r^2 = \pi(2.0 \times 10^{-2}\,\mathrm{m})^2 = 1.257 \times 10^{-3}\,\mathrm{m}^2$ (where $r = 2.0\,\mathrm{m}$ is the radius of the solenoid), to obtain $\mathrm{PE_M}/l$, the magnetic energy stored in the solenoid per unit length:

$$\begin{aligned}\frac{\mathrm{PE_M}}{l} &= \frac{\frac{1}{2}\mu_0 n^2 I^2 Al}{l} = \frac{1}{2}\mu_0 n^2 I^2 A \\ &= \frac{1}{2}(4\pi \times 10^{-7}\,\mathrm{T\cdot m/A})(2.0 \times 10^3/\mathrm{m})^2(5.0\,\mathrm{A})^2(1.257 \times 10^{-3}\,\mathrm{m}^2) \\ &= 7.9 \times 10^{-2}\,\mathrm{J/m} = 79\,\mathrm{mJ/m}\,.\end{aligned}$$

21 *AC and Electronics*

Answers to Selected Discussion Questions

•21.3

Unlike ac, dc does not induce currents in nearby conductors, thereby reducing the transmitted power. dc can be transmitted at higher voltages since it does not peak. The same rms voltage for an ac signal will rise to a peak voltage (1.414 times higher) and be more troublesome as far as insulation and breakdown of air is concerned).

•21.5

Across A-C, the scope reads the sinusoidal voltage induced on the output side of the transformer. Across A-B, the scope reads the voltage across the diode, which is zero when current is flowing through (when the diode acts like a closed switch) and equals the transformer output voltage when no current is flowing (when the diode acts like an open switch). Across B-C, the meter reads the voltage across the resistor, which equals the transformer output voltage when current is flowing and equals zero when no current is flowing.

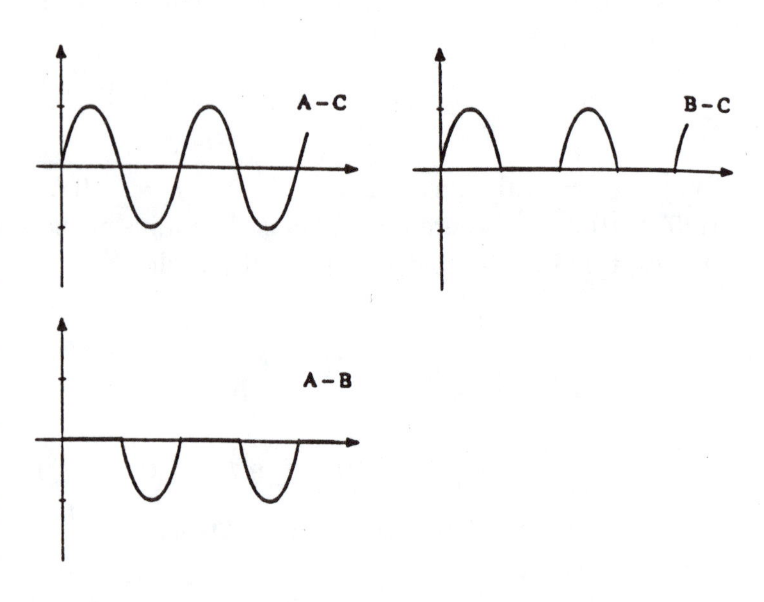

•21.7
Put the tester across the terminals of the fuse holder without the fuse in place — if all is well, the side coming into the house will be hot, the other will be floating, there will be a sizable potential difference and the neon will light. Now screw in the fuse and repeat the last test. If the fuse is good and if it is properly seated in the holder, there will be no appreciable voltage drop across it and the tester will not light. If it does, either the fuse is no good or it's not making good contact in the socket.

•21.9
When $f \to 0$ (that is, dc), the ratio is 1 (whatever low-frequency signals were in the input are for the most part in the output). This arrangement passes low frequencies on to the next circuit. When $f \to \infty$, the ratio approaches 0, little or no high-frequency voltage appears across C. At $f \approx 0$, the capacitor is open circuited. At $f \approx \infty$, the capacitor is short circuited.

•21.13
It's a one-transistor AM radio. The antenna feeds a tiny voltage (more or less, proportional to the length of the antenna wire) to the tuning circuit made up of the antenna coil and the variable capacitor. When tuned to resonate at the frequency of a particular station, current is maximum and the signal is passed on to the diode, which rectifies it (that is, it cuts off the negative portion). The transistor amplifies the positive signal so it will easily power the headphones, which respond to the envelope of the signal curve.

Answers to Odd-Numbered Multiple Choice Questions

1. b **3.** d **5.** e **7.** c **9.** b **11.** e **13.** a
15. b

Solutions to Problems

21.5
The *rms* (i.e., effective) voltage V_{rms} is related to the maximum voltage V_{m} via Eq. (21.8), so

$$V_{\mathrm{m}} = \sqrt{2}V_{\mathrm{rms}} = \sqrt{2}(100\,\mathrm{V}) = 141\,\mathrm{V}\,.$$

21.17

The fuse would not have blown if all the components were functioning properly. In fact the equivalent resistance R_e as indicated in Fig. P17 is $R_e = 5.0\,\Omega + 10\,\Omega + 50\,\Omega = 65\,\Omega$ which, with the 120-V power supply, results in a current of $I_{eff} = V_{eff}/R_e = 120\,\text{V}/65\,\Omega = 1.8\,\text{A}$, considerably below the limit of 3 A that would cause the fuse to blow. One of the three resistors is malfunctioning, probably shorted internally (meaning that it has an effective resistance of almost zero). This couldn't be the 5.0-Ω one or even the 10-Ω one; in fact even if *both* these smaller resistors are shorted there would still be 50 Ω left in the circuit, and the current would still be only $120\,\text{V}/50\,\Omega = 2.4\,\text{A}$, which remains under 3.0 A. So the one that is malfunctioning must be the 50-Ω one, which, when shorted, brings the effective resistance of the circuit down to $5.0\,\Omega + 10\,\Omega = 15\,\Omega$, and brings up the current in the circuit to $I = 120\,\text{V}/15\,\Omega = 8.0\,\text{A} > 3.0\,\text{A}$, causing the fuse to blow.

21.23

Within each period T the ac voltage $v(t)$ depicted in Fig. P23 is a linear function of time. In particular, the expression $v(t) = kt + b$ applies for $0 \le t \le T$. Here $b = -V_m$ is the intercept (i.e., the value of v at $t = 0$), while k is the slope. Since v rises from $-V_m$ to $+V_m$ over one period, $k = \text{rise/run} = [V_m - (-V_m)]/T = 2V_m/T$. Thus

$$v(t) = kt + b = \left(\frac{2V_m}{T}\right)t - V_m \qquad (0 \le t \le T)\,.$$

The average value of $v^2(t)$ over one period is then

$$\begin{aligned}
[v^2(t)]_{av} &= \frac{1}{T}\int_0^T v^2(t)\,dt = \frac{1}{T}\int_0^T \left[\left(\frac{2V_m}{T}\right)t - V_m\right]^2 dt \\
&= \frac{1}{T}\int_0^T \left(\frac{4V_m^2}{T^2}t^2 - \frac{4V_m^2}{T}t + V_m^2\right) dt \\
&= \frac{1}{T}\left[\frac{4V_m^2}{3T^2}t^3 - \frac{4V_m^2}{2T}t^2 + V_m^2 t\right]_0^T \\
&= \frac{V_m^2}{3}\,.
\end{aligned}$$

It follows that

$$V_{eff} = V_{rms} = \sqrt{[v^2(t)]_{av}} = \sqrt{\frac{V_m^2}{3}} = \frac{V_m}{\sqrt{3}}\,.$$

21.49

The reactance of a capacitor of capacitance C in an ac circuit with frequency f is given by Eq. (21.19): $X_C = 1/2\pi fC$. The current I drawn by such a reactance when it is connected to a power supply with an effective output voltage of V_{eff} is then given by

$$I_{eff} = \frac{V_{eff}}{X_C} = \frac{V_{eff}}{1/2\pi fC} = 2\pi fCV_{eff}\,.$$

Plug in $I_{\text{eff}} = 1.00\,\text{A}$, $V_{\text{eff}} = 120\,\text{V}$, $f = 60\,\text{Hz}$, and solve for C, the capacitance:

$$C = \frac{I_{\text{eff}}}{2\pi f V_{\text{eff}}} = \frac{1.00\,\text{A}}{2\pi(60\,\text{Hz})(120\,\text{V})} = 2.2 \times 10^{-5}\,\text{F} = 22\,\mu\text{F}\,.$$

21.75

The impedance Z of the circuit is a combination of its reactance X and resistance R: $Z = \sqrt{X^2 + R^2}$ [see Eq. (21.31)]. When connected to an ac generator with a terminal voltage of V_{eff}, the circuit draws an effective current I_{eff} which satisfies Eq. (21.32), $V_{\text{eff}} = I_{\text{eff}} Z$. Combine these two equations to obtain

$$V_{\text{eff}} = I_{\text{eff}}\sqrt{X^2 + R^2}\,.$$

Now plug in $V_{\text{eff}} = 60\,\text{V}$, $I_{\text{eff}} = 26\,\text{mA} = 0.026\,\text{A}$, and $X = 1.55\,\text{k}\Omega = 1.55 \times 10^3\,\Omega$, and solve for R:

$$R = \sqrt{\left(\frac{V_{\text{eff}}}{I_{\text{eff}}}\right)^2 - X^2} = \sqrt{\left(\frac{60\,\text{V}}{0.026\,\text{A}}\right)^2 - (1.55 \times 10^3\,\Omega)^2} = 1.7 \times 10^3\,\Omega = 1.7\,\text{k}\Omega\,.$$

21.81

The maximum current in an L-R-C circuit occurs when the frequency f of the circuit coincides with the resonance frequency f_0, which is given by Eq. (21.35):

$$f_0 = \frac{1}{2\pi\sqrt{LC}}\,.$$

The inductance which satisfies this formula is $L = 1/4\pi^2 f_0^2 C$. In our case $f_0 = 40.0\,\text{kHz} = 40.0 \times 10^3\,\text{Hz}$ and $C = 300\,\text{pF} = 300 \times 10^{-12}\,\text{F}$, so the inductance must be

$$L = \frac{1}{4\pi^2 f_0^2 C} = \frac{1}{4\pi^2(40.0 \times 10^3\,\text{Hz})^2(300 \times 10^{-12}\,\text{F})} = 5.28 \times 10^{-2}\,\text{H} = 52.8\,\text{mH}\,.$$

21.87

Suppose that the capacitance of the capacitor is C. Then in an ac circuit with frequency f the (capacitive) reactance of the capacitor is $X = X_C = 1/2\pi f C$, as per Eq. (21.19). Combine this with the resistive load R of the circuit to find the total impedance Z to be $Z = \sqrt{X^2 + R^2} = \sqrt{(1/2\pi f C)^2 + R^2}$ [see Eq. (21.31)]. If the effective voltage output of the power supply is V then from Ohm's Law [Eq. (21.32)] the *rms*, i.e., effective, current I must satisfy

$$V = IZ = I\sqrt{\left(\frac{1}{2\pi f C}\right)^2 + R^2}\,,$$

which we solve for C:

$$C = \frac{1}{2\pi f\sqrt{\left(\frac{V}{I}\right)^2 - R^2}} = \frac{1}{2\pi(60\,\mathrm{Hz})\sqrt{\left(\frac{120\,\mathrm{V}}{1.20\,\mathrm{A}}\right)^2 - (20\,\Omega)^2}} = 27\,\mu\mathrm{F}\,.$$

21.103

The voltages in the primary (p) and secondary (s) sides of a transformer are related via Eq. (21.36): $V_p/V_s = N_p/N_s$. In this case $N_p/N_s = 4/1$ and $V_p = 120\,\mathrm{V}$, so the secondary voltage is

$$V_s = V_p\left(\frac{N_s}{N_p}\right) = (120\,\mathrm{V})\left(\frac{1}{4}\right) = 30\,\mathrm{V}\,.$$

In an ideal transformer there is no loss of electromagnetic energy so Eq. (21.38) holds true: $I_pV_p = I_sV_s$. Solve for I_s, the secondary current:

$$I_s = \frac{V_pI_p}{V_s} = \frac{(120\,\mathrm{V})(1.0\,\mathrm{A})}{30\,\mathrm{V}} = 4.0\,\mathrm{A}\,.$$

21.107

By definition the efficiency e of a transformer is the ratio of output power vs the total input power. In this case the input power is $\mathrm{P}_{\mathrm{in}} = 45\,\mathrm{kW} + 300\,\mathrm{W} + 500\,\mathrm{W} = 45.8\,\mathrm{kW}$, of which 45 kW is the output power $\mathrm{P}_{\mathrm{out}}$ delivered to the secondary side. Thus

$$e = \frac{\mathrm{P}_{\mathrm{out}}}{\mathrm{P}_{\mathrm{in}}} = \frac{45\,\mathrm{kW}}{45.8\,\mathrm{kW}} = 0.983 = 98.3\%\,.$$

22 Radiant Energy: Light

Answers to Selected Discussion Questions

•22.3

A hair dryer radiates radiowaves that can be picked up as noise in the picture on a nearby TV set.

•22.7

The modern theory envisions light as both wave and particle; light energy is quantized but the propagation of those packets of energy is determined by its wave nature. Newton's version had a remarkably similar wave-particle structure: light was particulate but the particles were guided through space by wave patterns they set up in the aether.

•22.9

The radiation is absorbed by the moist meat because it contains water. The plate is dry and stays cool. The water molecules in an ice cube cannot undergo rotational motion and will not absorb microwaves until some of the surface has melted.

•22.13

A pulse can be a perfectly good wave. It's not obvious what is waving in a lightwave — it's certainly not a material aether. Later on, we'll talk about probability waves, but for the time being, let's say that the electromagnetic field itself waves.

•22.15

The sphere is in equilibrium with its weight down, balanced by an upward force exerted by the beam. Radiant energy can therefore transfer momentum and exert pressure. A sail craft for space travel is quite possible.

Answers to Odd-Numbered Multiple Choice Questions

1. c **3.** b **5.** a **7.** b **9.** a **11.** d **13.** c
15. d **17.** b **19.** e **21.** c

Solutions to Problems

22.7

Substitute $x = 0$ and $t = 0$ into the expression for E given in the problem statement:

$$\begin{aligned} E &= \left\{ (20\,\mathrm{V/m}) \cos \frac{2\pi}{1.00\,\mathrm{mm}} \left[x - (3.00 \times 10^8\,\mathrm{m/s})t\right] \right\}_{x=0,\, t=0} \\ &= (20\,\mathrm{V/m}) \cos \frac{2\pi}{1.00\,\mathrm{mm}} \left[0 - (3.00 \times 10^8\,\mathrm{m/s}) \times 0\right] \\ &= (20\,\mathrm{V/m}) \cos 0 = 20\,\mathrm{V/m}\,. \end{aligned}$$

22.13

To pick up a signal broadcasted at frequency f the resonance frequency f_0 of the tuning L-C circuit in the radio must be adjusted to match f, i.e.,

$$f = f_0 = \frac{1}{2\pi\sqrt{LC}}\,.$$

Solve for L:

$$L = \frac{1}{4\pi^2 f^2 C} = \frac{1}{4\pi^2 (100 \times 10^6\,\mathrm{Hz})^2 (0.5 \times 10^{-12}\,\mathrm{F})} = 5 \times 10^{-6}\,\mathrm{H} = 5\,\mu\mathrm{H}\,.$$

22.19

Start with the expression for the wave in Problem (22.10): $E = E_0 \sin k(x - v t) = E_0 \sin(kx - kvt)$. Noting that $k = 2\pi/\lambda$ and $v = \lambda/T = \lambda f$, we have

$$kv = \left(\frac{2\pi}{\lambda}\right)(\lambda f) = 2\pi f = \omega\,,$$

whereby

$$E = E_0 \sin(kx - kvt) = E_0 \sin(kx - \omega t)\,.$$

22.23

Compare the expression given in the problem statement, $E = 2.0 \times 10^2 \sin[3.0 \times 10^6 \pi(x - 3.0 \times 10^8 t)]$, with the standard form given in Eq. (22.5): $E = E_0 \sin[(2\pi/\lambda)(x - vt)]$, to obtain the amplitude of the E-component of the TEM wave to be $E_0 = 2.0 \times 10^2\ \mathrm{V/m}$. Now apply Eq. (22.2), $E = \mathrm{c}B$, to find the amplitude of the B-component to be

$$B_0 = \frac{E_0}{\mathrm{c}} = \frac{2.0 \times 10^2\ \mathrm{V/m}}{3.0 \times 10^8\ \mathrm{m/s}} = 6.7 \times 10^{-7}\ \mathrm{T}\,.$$

22.61

The kinetic energy of the electron after being accelerated through an electric potential difference ΔV is $\mathrm{KE} = q_e \Delta V$, where $q_e = 1.602 \times 10^{-19}\ \mathrm{C}$ is the magnitude of one electron charge. If this much energy is entirely absorbed by one single atom which subsequently emits an X-ray photon, then according to the Conservation of Energy the maximum energy E_{max} carried by the photon must be equal to the kinetic energy of the electron prior to its collision with the atom: $\mathrm{E}_{max} = \mathrm{KE} = q_e \Delta V$. Plug in $\Delta V = 1.00 \times 10^4\ \mathrm{V}$ to obtain

$$\mathrm{E}_{max} = q_e \Delta V = q_e (1.00 \times 10^4\ \mathrm{V}) = 1.00 \times 10^4 q_e \cdot \mathrm{V} = 1.00 \times 10^4\ \mathrm{eV}\,,$$

where we used the definition of eV: $1\ \mathrm{eV} = 1\, q_e \cdot \mathrm{V}$.

Since the wavelength λ of a photon is related to its energy E as $\mathrm{E} = hf = hc/\lambda$, λ reaches its minimum value when $\mathrm{E} = \mathrm{E}_{max}$:

$$\lambda_{min} = \frac{hc}{\mathrm{E}_{max}} = \frac{(4.136 \times 10^{-15}\ \mathrm{eV{\cdot}s})(2.998 \times 10^8\ \mathrm{m/s})}{1.00 \times 10^4\ \mathrm{eV}} = 1.24 \times 10^{-10}\ \mathrm{m} = 0.124\ \mathrm{nm}\,.$$

23 *The Propagation of Light: Scattering*

Answers to Selected Discussion Questions

•23.1
(a) It's mostly diffuse, although there is a little specular. (b) Flat paints have a diffuse, frequency-independent surface reflection that results in a white haze. (c) The index of water is between that of the fibers and the air so there is less diffuse white light reflected when it's wet. There will be a whitish haze over the dry painting.

•23.7
It seems that the mirror is too far toward her middle for her to be looking at herself. She's looking at you which makes the picture strangely interesting.

•23.9
The gentleman is standing where we are and her image is shifted too far to the right to be produced by a flat mirror parallel to the bar.

•23.11
The resonance in the UV in part accounts for the effective absorption of UV by glass, which is essentially opaque to it. Only in great thicknesses will the weak absorption of red and blue produce a greenish tint.

•23.13
(a) Red. (b) C = W − R, cyan ink "eats" red; hence, it will appear black. (c) It absorbs blue and produces more contrast between sky and cloud.

•23.17

(a) C = B+G — the filter absorbs B; hence, only G emerges. (b) Y = R+G — the filter passes R+B and absorbs G; hence, R is transmitted. (c) (B+G)+(R+B) = (R+B+G)+B = W+B, unsaturated blue. (d) (R + G + B) − R − G = B.

Answers to Odd-Numbered Multiple Choice Questions

1. c **3.** d **5.** e **7.** a **9.** d **11.** d **13.** d
15. d **17.** c **19.** e **21.** d

Solutions to Selected Problems

23.5

The incident light rays are almost parallel to the surface of the mirror, so they make an angle of nearly 90° with respect to the normal to the surface. By definition this is the angle of incidence; i.e., $\theta_i \approx 90°$. According to the Law of Reflection [see Eq. (23.1)] the angle of reflection θ_r is $\theta_r = \theta_i \approx 90°$ as well.

23.11

If the woman is at a distance s from the mirror then her image appears to be a distance s behind the mirror, for a total of $s + s = 2s$ from herself. Since she can see things clearly from a distance of 25.0 cm we set $2s = 25.0\,\text{cm}$ to obtain $s = 12.5\,\text{cm}$, i.e., the mirror should be placed a distance 12.5 cm in front of the woman for her to see her own image clearly.

23.21

The case when the length H of the mirror is at its minimum value that will suffice is depicted in Fig. P21 in the text. The image of the chart in the mirror is $\overline{C_1'C_2'}$, which by mirror symmetry also has a height h and is equidistant from the mirror with the chart itself. Now consider the triangle ΔOM_1M_2, with a base length of $\overline{M_1M_2} = H$ and height d (which is the vertical distance from vertex O to the baseline $\overline{M_1M_2}$); and the larger triangle $\Delta OC_1'C_2'$, with a corresponding base length of $\overline{C_1'C_2'} = h$ and height $s_o + d$. Similarity between the two triangles requires that $H/d = h/(s_o + d)$, which we solve for H, the minimum length of the mirror:

$$H = \frac{hd}{s_o + d}.$$

23.29

The speed of light in a medium is given by Eq. (23.2), $v = c/n$, where c is the speed of light in vacuum and n is the index of refraction of the medium. In the case of diamond $n = 2.42$, so the speed of light in diamond is

$$v = \frac{c}{n} = \frac{2.998 \times 10^8\,\text{m/s}}{2.42} = 1.24 \times 10^8\,\text{m/s}\,.$$

23.45

Consider an arbitrary point O, a distance h below the surface of the water. Imagine two light rays, labeled 1 and 2, respectively, that emerge from O into the air. Ray 1 is perpendicular to the air-water boundary and makes its way into the air undeflected, while ray 2 is incident upon the boundary at an angle θ_i and emerges into the air at an angle θ_t, as shown. Viewed from above in the air the apparent position of point O is O', which appears to be the location where rays 1 and 2 would originate had they not been deflected. Thus the apparent depth of point O is h', which is obviously shallower than h, the true depth.

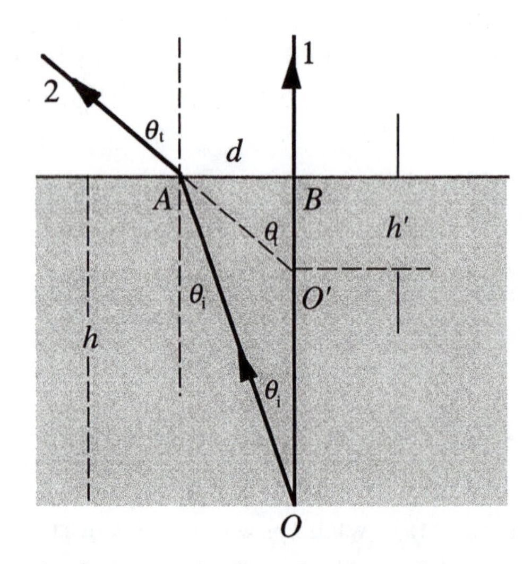

To find h' in terms of h, apply the Law of Refraction, $n_i \sin\theta_i = n_t \sin\theta_t$, where in this case $n_i = 1.333$ for water and $n_t = 1.00$ for air, so $\sin\theta_i = (n_t/n_i)\sin\theta_t = (1.00/1.333)\sin\theta_t = \frac{3}{4}\sin\theta_t$. If we view the water almost perpendicularly then both θ_i and θ_t are very small, in which case $\sin\theta_i \approx \tan\theta_i$ and $\sin\theta_i \approx \tan\theta_i$. Thus θ_i is related to θ_t by

$$\tan\theta_i \approx \frac{3}{4}\tan\theta_t\,.$$

Now consider the right-angled triangles $\triangle OAB$ and $\triangle O'AB$, in which $d/h' = \tan\theta_t$ and $d/h = \tan\theta_i$. Solve for h' and h to obtain $h' = d/\tan\theta_t$ and $h = d/\tan\theta_i$. Divide these two equations to obtain

$$\frac{h'}{h} = \frac{d/\tan\theta_t}{d/\tan\theta_i} = \frac{\tan\theta_i}{\tan\theta_t} = \frac{3}{4}\,,$$

where we used the relation between $\tan\theta_i$ and $\tan\theta_t$ we derived earlier. This proves that $h' = \frac{3}{4}h$.

23.63

Label the material whose index of refraction is 1.33 as material 1 and the other one as material 2. Since $n_1 = 1.33 < n_2 = 2.40$, total internal reflection can only occur when a beam of light is incident from material 2 onto the interface between the two materials. The minimum angle of incidence for this to occur is the critical angle θ_c, given by Eq. (23.5): $\sin\theta_c = n_t/n_i$. In this case $n_t = n_1 = 1.33$ and $n_i = n_2 = 2.40$, so

$$\sin\theta_c = \frac{n_t}{n_i} = \frac{n_1}{n_2} = \frac{1.33}{2.40} = 0.554\,17\,,$$

which gives $\theta_c = \sin^{-1} 0.554\,17 = 33.7°$.

23.67

Since $n_{\text{air}} < n_{\text{water}}$ the angle of incidence of a light ray entering the water surface from the air is greater than the angle of transmission: $\theta_i > \theta_t$. As we gradually look away from point B (which is directly on top of the fish at point c) towards point A, θ_i gradually increases. At point A, the edge of what the fish can see, θ_i has reached 90°, its greatest possible value, and the incoming light rays arriving at point A are essentially parallel to the surface of the water. The fish can only see the outside world through a circular portion of the water surface, centered at point B with a radius $\overline{AB}$.

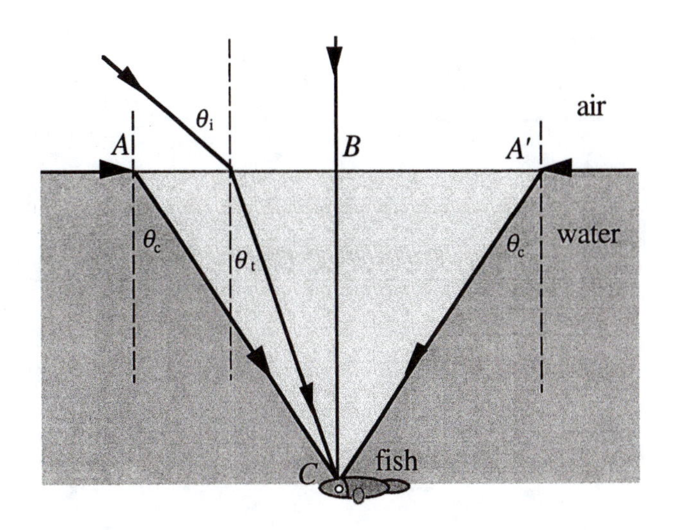

Now plug $\theta_i = 90°$ into Eq. (23.3), along with $n_i = n_{\text{air}} = 1.00$ and $n_t = n_{\text{water}} = 1.333$, and solve for θ_t, the corresponding angle of transmission:

$$\sin\theta_t = \frac{n_{\text{air}}\sin\theta_i}{n_{\text{water}}} = \frac{1.00(\sin 90°)}{1.333} = 0.750\,2\,,$$

which gives $\theta_t = \sin^{-1} 0.750\,2 = 48.6°$. This is in fact the critical angle for total internal reflection to occur at the water-air surface, satisfying $\sin\theta_c = n_{\text{air}}/n_{\text{water}}$. The cone angle is therefore $2\theta_c = 2(48.6°) = 97.2°$.

24 *Geometrical Optics & ...*

Answers to Selected Discussion Questions

•24.1
The focal length increases because the rays are not bent as strongly at the water-glass interface.

•24.3
The focal length depends on the index of refraction and that depends on the wavelength.

•24.5
Form a real image of a very distant object; the image-distance then approaches the focal length. Place the two in contact, shine in parallel light and measure f, knowing that $1/f = 1/f_+ + 1/f_-$, where f_+ is given and f_- is to be found.

•24.9
The radius of curvature is ∞ and so is f. That means the object- and image-distances must have equal magnitudes. Thus, the magnification is $+1$.

•24.13
The target is at one of the two focii of the hyperboloid and rays reflected from it appear to come from the other focus, $F_1(H)$, but this is also a focus, $F_1(E)$, of the ellipsoidal mirror. Rays appearing to come from one focus of the ellipsoid, after reflecting off it, converge to the other focus, $F_2(E)$, at the film plane.

•24.15

The object has a diameter d, where $d(1200) = 0.0005\,\text{m}$, $d = 4.2 \times 10^{-7}\,\text{m}$, which is the same as the wavelength of violet light. We cannot hope to see objects that are smaller than the probe being used; namely, the wavelength of light. The amount of diffraction will then obscure the image completely. So the details we wish to observe cannot be finer than λ, which puts a practical limit on the magnification. A 12000× microscope will only make larger blurred images showing no more detail.

•24.17

The filament is located at one focus of the ellipsoid and the reflected light is made to converge at the other focus, which also corresponds to the front focal point of the lens combination.

Answers to Odd-Numbered Multiple Choice Questions

1. b **3.** b **5.** d **7.** b **9.** a **11.** e **13.** b
15. d **17.** a **19.** e

Solutions to Problems

24.29

When an object is placed a distance s_o in front of a convex lens of focal length f a real image is formed a distance s_i behind the lens (if $s_o > f$). Here s_o, s_i and f are related via the Gaussian Lens Equation [Eq. (24.6)]: $1/f = 1/s_o + 1/s_i$. In our case the object (grasshopper) is located 10 cm to the left of the lens while its image is 30 cm to the right of the lens, so $s_o = 10\,\text{cm}$ and $s_i = 30\,\text{cm}$. Plug these data into Eq. (24.6) to find f, the focal length of the lens:

$$\frac{1}{f} = \frac{1}{s_o} + \frac{1}{s_i} = \frac{1}{10\,\text{cm}} + \frac{1}{30\,\text{cm}} = \frac{4}{30\,\text{cm}},$$

and so $f = 30\,\text{cm}/4 = 7.5\,\text{cm}$.

(a) Now the grasshopper jumps 7.5 cm towards the lens so s_o decreases by 7.5 cm, to $10\,\text{cm} - 7.5\,\text{cm} = 2.5\,\text{cm}$. Plug this new value of s_o, along with $f = 7.5\,\text{cm}$, into Eq. (24.6) again and solve for s_i, the new location of the image:

$$s_i = \frac{s_o f}{s_o - f} = \frac{(2.5\,\text{cm})(7.5\,\text{cm})}{2.5\,\text{cm} - 7.5\,\text{cm}} = -3.8\,\text{cm},$$

which means that the image is now 3.8 cm to the left of the lens (i.e., on the same side of the lens as the grasshopper itself), since $s_i < 0$.

(b) At first $s_i = 30\,\text{cm}$ and $s_o = 10\,\text{cm}$, so the transverse magnification is $M_T = -s_i/s_o = -30\,\text{cm}/10\,\text{cm} = -3.0$. The image is real (as $s_i > 0$), inverted (as $M_T < 0$), and magnified to three times the original size of the grasshopper (as $|M_T| = 3.0$).

Similarly, for the new image $s_i = -3.8\,\text{cm} < 0$, so it is now virtual. The transverse magnification is now $M_T = -s_i/s_o = -3.8\,\text{cm}/2.5\,\text{cm} = +1.5 > 0$, so the new image is right-side-up and is magnified to 1.5 times the original size.

(c) The two ray diagrams below depict the situation before and after the grasshopper makes the jump.

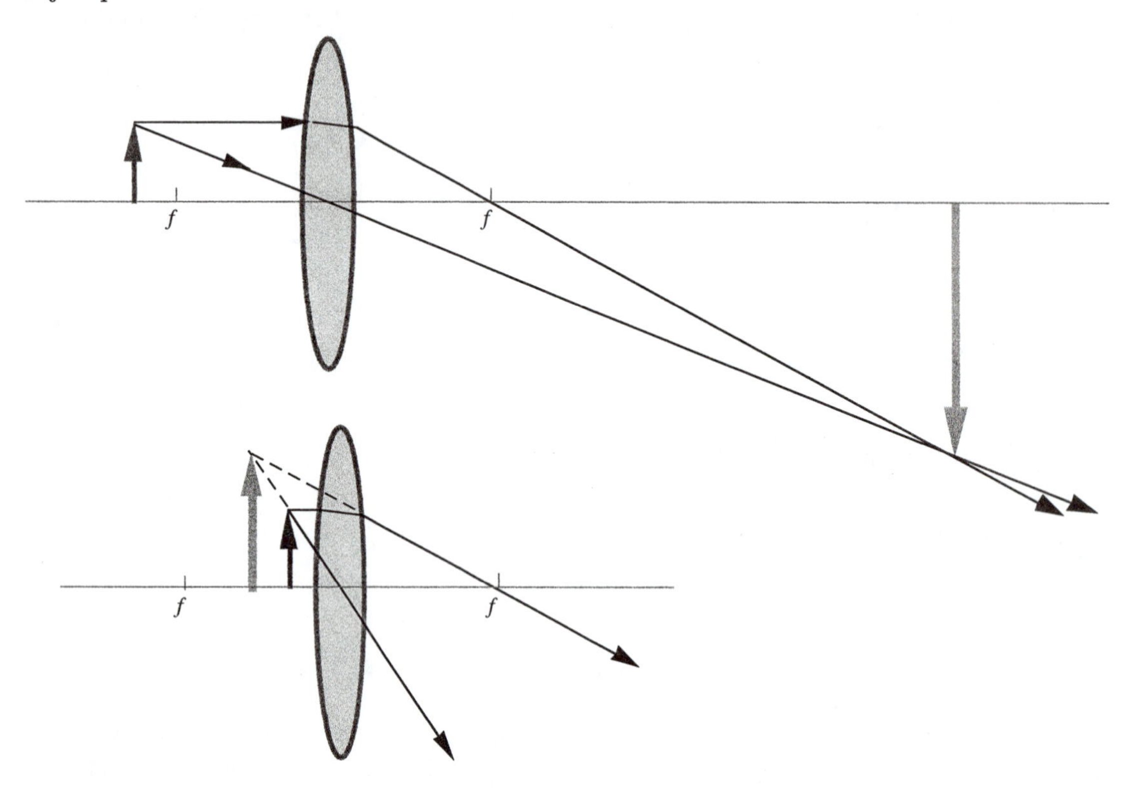

24.33

Consider the two similar triangles in Fig. P33, one with a base y_o and corresponding height s_o, and the other with a base y_i and corresponding height s_i. By similarity $|y_o/s_o| = |y_i/s_i|$, or $|y_i/y_o| = |s_i/s_o|$, so the transverse magnification of the pinhole camera is, by definition,

$$M_T = \frac{y_i}{y_o} = -\frac{s_i}{s_o},$$

which is identical with the expression for M_T for a lens. Note the negative sign we added here to reflect the fact that the image is inverted. This formula tells us that the magnification of the pinhole camera is proportional to s_i, the pinhole-film plane distance, which is about the same as the length of the camera box. To obtain a larger image, just extend the length of the camera body accordingly. (Many commercial pinhole cameras do come with an adjustable body length.)

24.37

The positive lens (with focal length $f = 60\,\mathrm{cm}$) projects the object, which is the picture on the TV screen, onto the wall, where the image of the TV screen is located. So the distance d between the TV screen and the wall is that between the object and the image: $d = s_\mathrm{o} + s_\mathrm{i}$.

Now let's find s_o and s_i. Since the picture is enlarged 3 times, the transverse magnification is [see Eq. (24.9)] $M_\mathrm{T} = -s_\mathrm{i}/s_\mathrm{o} = -3$. Here the minus sign corresponds to the fact that, being real, the image on the wall must inverted. The last equation can be rewritten as $s_\mathrm{i} = 3s_\mathrm{o}$, which we plug into Eq. (24.6), along with $f = 60\,\mathrm{cm}$:

$$\frac{1}{f} = \frac{1}{0.60\,\mathrm{m}} = \frac{1}{s_\mathrm{o}} + \frac{1}{s_\mathrm{i}} = \frac{1}{s_\mathrm{o}} + \frac{1}{3s_\mathrm{o}} = \frac{4}{3s_\mathrm{o}}\,,$$

which gives $s_\mathrm{o} = 4(0.60\,\mathrm{m})/3 = 0.80\,\mathrm{m}$. Thus $s_\mathrm{i} = 3s_\mathrm{o} = 3(0.80\,\mathrm{m}) = 2.4\,\mathrm{m}$. The distance between the TV screen and the wall is then

$$d = s_\mathrm{o} + s_\mathrm{i} = 0.80\,\mathrm{m} + 2.4\,\mathrm{m} = 3.2\,\mathrm{m}\,.$$

The primary benefit of using a large lens is that more light from the TV screen can be collected by the lens and projected onto the wall, resulting in a brighter image. Since the image on the wall is inverted (with respect to the picture on the TV screen), the TV set has to be placed upside down, if the final image on the wall is to appear normal (i.e., right-side-up).

24.53

By the definition of the tube length, the total separation L_total between the objective lens and the eyepiece is the sum of of the focal lengths f_O and f_E of the two lenses plus the tube length L:

$$L_\mathrm{total} = f_\mathrm{O} + f_\mathrm{E} + L\,.$$

In this case the total separation is $L_\mathrm{total} = 10.0\,\mathrm{cm}$, $f_\mathrm{O} = 10\,\mathrm{mm} = 1.0\,\mathrm{cm}$, and $f_\mathrm{E} = 30\,\mathrm{mm} = 3.0\,\mathrm{cm}$; so the tube length is

$$L = L_\mathrm{total} - f_\mathrm{O} - f_\mathrm{E} = 10.0\,\mathrm{cm} - 1.0\,\mathrm{cm} - 3.0\,\mathrm{cm} = 6.0\,\mathrm{cm}\,.$$

24.71

(a) We are given $s_\mathrm{o} = 30\,\mathrm{cm} = 0.30\,\mathrm{m}$ for the location of the object and $s_\mathrm{i} = 9.0\,\mathrm{m}$ for the location of the image. Note that here $s_\mathrm{i} > 0$ since the image is projected onto a screen, which must be on the same side of the mirror as the object — no light rays from the object can reach behind the mirror. These data lead to R, the radius of curvature of the mirror, via Eq. (24.21): $1/s_o + 1/s_i = -2/R$, or

$$R = -\frac{2}{1/s_o + 1/s_i} = -\frac{2}{1/0.30\,\text{m} + 1/9.0\,\text{m}} = -0.58\,\text{m}\,.$$

Note that $R < 0$ since the mirror is concave.

(b) The transverse magnification is given by Eqs. (24.8) and (24.9) as $M_T = y_i/y_o = -s_i/s_o$, where y_i and y_o are the transverse sizes of the image and the object, respectively. To find y_i, plug $y_o = 5.0\,\text{cm} = 0.050\,\text{m}$, along with $s_i = 9.0\,\text{m}$ and $s_o = 0.30\,\text{m}$, into the formula for M_T above and solve for y_i:

$$y_i = -\frac{y_o s_i}{s_o} = -\frac{(0.050\,\text{m})(9.0\,\text{m})}{0.30\,\text{m}} = -1.5\,\text{m}\,.$$

The image is therefore 1.5 m tall and inverted (as indicated by the minus sign in front of y_i). It is of courses real, since it can be projected onto a screen.

24.77

First we need to locate the image of the satellite formed by the mirror, by solving for s_i from Eq. (24.21), $1/s_o + 1/s_i = -2/R$. Assuming that the satellite is directly above in the sky, then s_o is approximately the altitude of its orbital: $s_o \approx 500\,\text{km} = 5.00 \times 10^5\,\text{m}$. Plug this value for s_o, along with $R = -1.0\,\text{m}$ (negative as the mirror is concave), into the equation above and solve for s_i:

$$s_i = -\frac{1}{2/R + 1/s_o} \approx -\frac{1}{2/(-1.0\,\text{m}) + 1/(5.00 \times 10^5\,\text{m})} = +0.50\,\text{m}\,,$$

so the (real) image of the satellite is located 0.50 m in front of the vertex of the mirror. The size of the image can then be obtained from the transverse magnification M_T in Eqs. (24.8) and (24.9) (which are identical for both thin lenses and spherical mirrors): $M_T = y_o/y_i = -s_i/s_o$. Here $y_o = 2.0\,\text{m}$ is the size of the satellite, $s_i = 0.50\,\text{m}$, $s_o = 5.00 \times 10^5\,\text{m}$, which we plug into the equation and solve for $|y_i|$, the size of the image of the satellite in the telescope:

$$|y_i| = \left|-\frac{y_o s_i}{s_o}\right| = \left|-\frac{(2.0\,\text{m})(0.50\,\text{m})}{5.00 \times 10^5\,\text{m}}\right| = 2.0 \times 10^{-6}\,\text{m} = 2.0\,\mu\text{m}\,.$$

(If you left out the absolute value sign you would obtain $y_i = -2.0\,\mu\text{m}$, where minus sign only indicates that the image is inverted.)

24.85

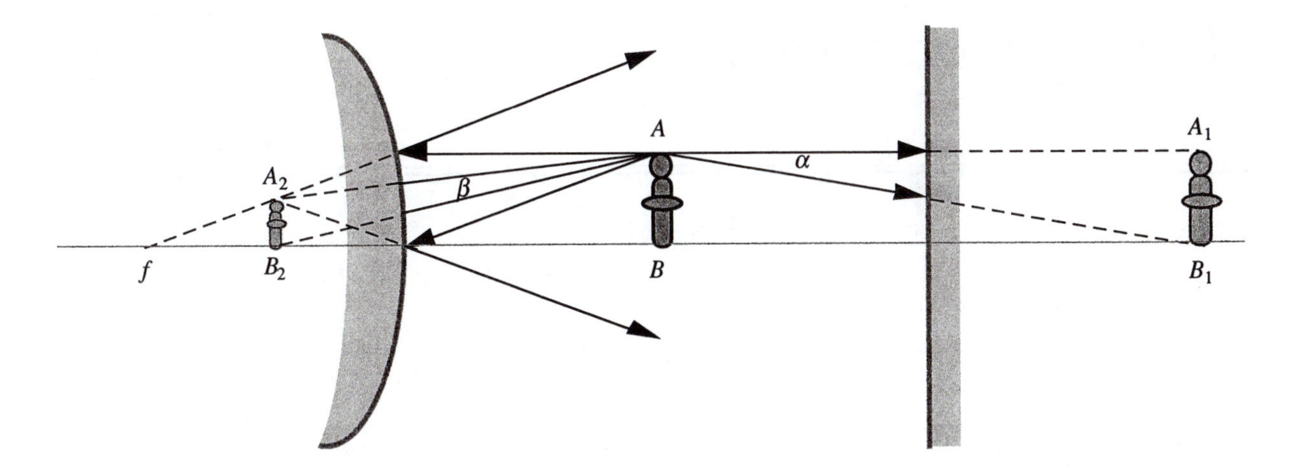

In the ray diagram above, the youngster (represented by $\overline{AB}$) sees her image in the plane mirror ($\overline{A_1B_1}$) twice as large as her image in the convex mirror ($\overline{A_2B_2}$). This means that $\alpha = 2\beta$, where α and β are the angles subtended by the images in the plane and convex mirrors, respectively, as shown. Since the youngster's size (1.00 m) is considerably less than the distance between her and her images, both α and β are small, so $\alpha \approx \overline{A_1B_1}/\overline{AA_1}$ and $\beta \approx \overline{A_2B_2}/\overline{AA_2}$, where $\overline{A_1B_1} = \overline{AB} = 1.0\,\text{m}$ (for plane mirror), $\overline{AA_1}$ is twice the the separation between her and the plane mirror, or $2 \times 5.0\,\text{m} = 10\,\text{m}$. These relationships can be combined to yield

$$\beta \approx \frac{\overline{A_2B_2}}{\overline{AA_2}} = \frac{\alpha}{2} \approx \frac{\overline{A_1B_1}}{2\overline{AA_1}} \approx \frac{1.0\,\text{m}}{2(10\,\text{m})} = 0.050\,,$$

i.e.,

$$\frac{\overline{A_2B_2}}{\overline{AA_2}} \approx 0.050\,. \tag{1}$$

Now consider the image in the convex mirror. We have $\overline{AA_2} \approx s_o + |s_i| = s_o - s_i$, where $s_o = 10\,\text{m}/2 = 5.0\,\text{m}$. Here we noted that $s_i < 0$ since the image is behind the vertex of the mirror. Thus Eq. (24.21) reads

$$\frac{1}{s_o} + \frac{1}{s_i} = \frac{1}{5.0\,\text{m}} + \frac{1}{5.0\,\text{m} - \overline{AA_2}} = \frac{1}{f}\,. \tag{2}$$

Finally, the magnification of the image in the convex mirror is

$$M_T = \frac{\overline{A_2B_2}}{\overline{AB}} = \frac{\overline{A_2B_2}}{1.0\,\text{m}} = -\frac{s_i}{s_o} = -\frac{5.0\,\text{m} - \overline{AA_2}}{5.0\,\text{m}}\,. \tag{3}$$

There are three variables, namely $\overline{A_2B_2}$, $\overline{AA_2}$, and f, in Eqs. (1) through (3) above. Solve them to obtain $\overline{A_2B_2} = 0.33\,\text{m}$, $\overline{AA_2} = 6.7\,\text{m}$, and $f = -2.5\,\text{m}$.

25 *Physical Optics*

Answers to Selected Discussion Questions

•25.1

Yes. Monochromatic light can be imagined as the superposition of many waves with different polarization states but all of the same wavelength and all infinitely long. There cannot be any random phase changes because each component is infinitely long and constant in phase. So all will simply combine to form a constant polarization of some sort.

•25.5

Since the index of refraction of benzene is around 1.5, the polarization angle will increase and the reflection again become visible.

•25.7

W = R+B+G; red is strongly transmitted so the light passed is unsaturated red, and B+G = C is reflected.

•25.9

The speckle effect arises from the interference of light reflected from adjacent regions of the surface. The more coherent the illuminating light, the more apparent the phenomenon. With a laser the speckles almost fill the space in front of the illuminated surface.

•25.11

For two holes on a horizontal line the pattern is as shown here. There will be a series of Young's fringes within the diffraction pattern of the individual apertures; namely, the Airy pattern.

•25.17

(a) 5; (b) 6; (c) 3; (d) 2; (e) 4; (f) 1; (g) 7.

Answers to Odd-Numbered Multiple Choice Questions

1. c **3.** a **5.** b **7.** a **9.** c **11.** c **13.** b
15. a

Solutions to Problems

25.9

The incident beam is polarized in the vertical direction while the axis of polarization of the polarizer is at 30° above the horizontal, so the angle between the two is $\theta = 90° - 30° = 60°$. The transmitted irradiance I can then be obtained from Malus's Law, Eq. (25.1): $I = I_1 \cos^2\theta$, with $I_1 = 160\,\mathrm{W/m^2}$ the irradiance of the incident beam:

$$I = I_1 \cos^2\theta = (160\,\mathrm{W/m^2})(\cos^2 60°) = 40\,\mathrm{W/m^2}\,.$$

25.11

Denote the irradiance of the incident light beam as I_0. The angle θ_1 between the direction of polarization of the incident beam and the axis of the first polarizer is $\theta_1 = 40° - 10° = 30°$, so the irradiance of the light transmitted through the first polarizer is $I_1 = I_0 \cos^2\theta_1$. Now, the angle θ_2 between the axes of the two polarizers is $\theta_2 = 70° - 10° = 60°$, so the final irradiance I of the light emerging from the second polarizer is [see Eq. (25.1)]

$$I = I_1 \cos^2\theta_2 = (I_0 \cos^2\theta_1)\cos^2\theta_2 \,.$$

The fraction of the incoming light that emerges from the second polarizer is therefore

$$\frac{I}{I_0} = \frac{I_0 \cos^2\theta_1 \cos^2\theta_2}{I_0} = \cos^2\theta_1 \cos^2\theta_2 = \left(\cos^2 30\right)\left(\cos^2 60\right) = 0.19 \,.$$

25.47

In the diagram shown to the right the two speakers are denoted as S_1 and S_2. If the listener is located at point C, on the central axis bisecting $\overline{S_1S_2}$, then $r_1 = r_2$, meaning that no additional phase difference arises between the signals from S_1 and S_2 en route to point C — the phase difference between the two signals arriving at that point is entirely due to their initial difference. Since they are initially out-of-phase, they will arrive at point C still out-of-phase. So the listener at that point encounters an interference minimum — certainly not a good spot to enjoy the music.

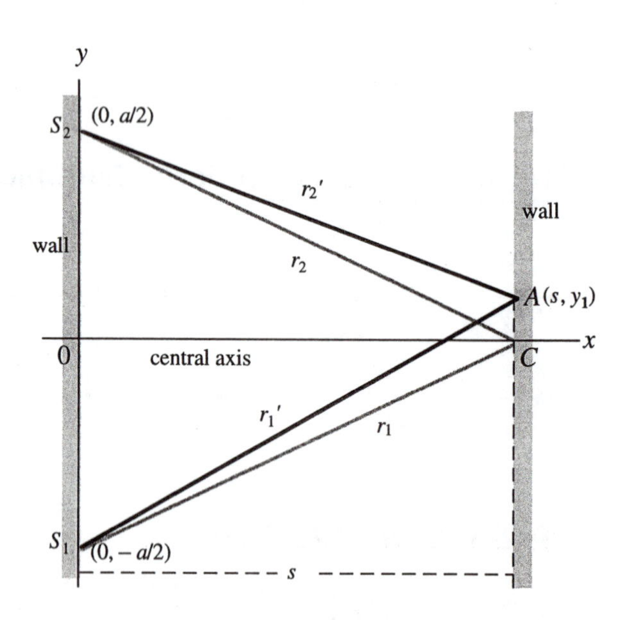

Now suppose the listener moves along the wall to point A, which is the normal location of the first-order interference maximum had there been no initial phase difference between the waves from both speakers, then $r_1' - r_2' = m\frac{1}{2}\lambda$, with $m = 1$ and λ the wavelength of the sound wave from the speakers. The net phase difference between the two waves at A, due to both the initial phase difference ($\pm\pi$) and the path length difference (π), would now be $\pm\pi + \pi = 0$ or 2π, which would make A the location of an interference maximum. (Here the $\pm$ sign depends on whether the initial phase of the wave from S_1 leads or trials that from S_2, and it does not affect the result.)

Since $a = \overline{S_1S_2} = 5.00\,\text{m}$ and $s = 10.0\,\text{m}$, the condition $a \ll s$ is not valid so we cannot expect to approximation in the text, namely $r_1' - r_2' \approx ay/s$, to be valid to a great deal of accuracy. To find a better approximation of $r_1' - r_2'$, note from the figure that $r_1' = \sqrt{s^2 + (a/2 + y_1)^2}$ and $r_2' = \sqrt{s^2 + (a/2 - y_1)^2}$. Suppose that $\overline{AC} = y_1 \ll s$ and a, then

$$r_1' = \sqrt{s^2 + \frac{a^2}{4} + ay_1 + y_1^2} \approx \sqrt{s^2 + \frac{a^2}{4} + ay_1} \,.$$

Using the binomial approximation $\sqrt{1+x} \approx 1 + \frac{1}{2}x$ for $|x| \ll 1$, this can be further simplified to be

$$r_1' \approx \sqrt{s^2 + \frac{a^2}{4}} \times \sqrt{1 + \frac{ay_1}{s^2 + a^2/4}} \approx \sqrt{s^2 + \frac{a^2}{4}} \left(1 + \frac{1}{2}\frac{ay_1}{s^2 + a^2/4}\right)$$
$$= \sqrt{s^2 + \frac{a^2}{4}} + \frac{1}{2}\frac{ay_1}{\sqrt{s^2 + a^2/4}} .$$

Similarly, for r_2' we have $r_2' \approx \sqrt{s^2 + a^2/4} - \frac{1}{2}\, ay_1/\sqrt{s^2 + a^2/4}$; and so

$$r_1' - r_2' \approx \frac{ay_1}{\sqrt{s^2 + a^2/4}} .$$

Equate this with $\frac{1}{2}\lambda$, with $\lambda = v/f$ (where $v = 346\,\text{m/s}$ and $f = 1000\,\text{Hz}$), and solve for y_1:

$$y_1 \approx \frac{mv\sqrt{s^2 + a^2/4}}{2fa} = \frac{(1)(346\,\text{m/s})\sqrt{(10.0\,\text{m})^2 + (5.00\,\text{m})^2/4}}{2(1000\,\text{Hz})(5.00\,\text{m})} = 0.357\,\text{m} .$$

So the listener has to move by 0.357 m to hear the first maximum in sound intensity from the two speakers. (Note that, if you applied the approximation $y_1 \approx ay_1/s$ by ignoring the difference between s and $\sqrt{s^2 + a^2/4}$, you would get $y_1 \approx 0.346\,\text{m}$. Also, the value of y_1 turns out to be indeed much less than either s or a, so our approximation is self-consistent.)

25.55

The value of the index of refraction of the film is $n_f = n_e = 1.36$, which is in between those of n_1 (the air, $= 1.00$) and n_2 (the glass): $n_1 < n_f < n_2$. Thus the wavelength λ_0 of the light which results in maximum reflection is given by Eq. (25.11): $d = m\lambda_0/2n_f$, where d is the thickness of the film and $m = 1, 2, 3, \ldots$. Plug in $\lambda_0 = 500\,\text{nm}$ (for green light) to obtain

$$d = \frac{m\lambda_0}{2n_f} = \frac{m(500\,\text{nm})}{2(1.36)} = m(184\,\text{nm}) . \qquad (m = 1, 2, 3, \ldots)$$

For the minimum value of d, set $m = 1$: $d_{\text{min}} = 184\,\text{nm} = 1.84 \times 10^{-7}\,\text{m}$.

We can easily verify that, at this thickness, no other wavelength in the visible light range will result in maximum reflection. In fact the wavelength which satisfies the condition for maximum reflection above is $\lambda_0 = 2n_f d_{\text{min}}/m = 2(1.36)(184\,\text{nm})/m = 500\,\text{nm}/m$. While $m = 1$ gives the desired green color, $m = 2$ would mean $\lambda_0 = 250\,\text{nm}$, which is already out of the visible light range ($390\,\text{nm} < \lambda_0 < 780\,\text{nm}$). Any greater values of m would only make λ_0 even shorter.

25.77

The central maximum in a single-slit diffraction pattern is bounded by the two first-order minima, which satisfy Eq. (25.13), $D\sin\theta_{m'} = m'\lambda$, with $m' = \pm 1$. Here D is the width of the slit and λ is the wavelength. The angular displacement θ_1 of the irradiation minimum corresponding to $m' = +1$, measured from the central axis, can therefore be obtained from $\sin\theta_1 = \lambda/D$. Plug in $\lambda = 461.9\,\text{nm}$ and $D = 0.10\,\text{mm} = 0,10 \times 10^6\,\text{nm}$ to obtain

$$\sin\theta_1 = \frac{\lambda}{D} = \frac{461.9\,\text{nm}}{0.10 \times 10^6\,\text{nm}} = 4.619 \times 10^{-3},$$

which yields $\theta_1 = \sin^{-1}(4.619 \times 10^{-3}) = 0.264\,65°$. This is the angular width between the central axis and the first minimum on one side. Due to the symmetry of the diffraction pattern about the central axis the angular width of the entire central peak is twice that much, or $2\theta_1 = 2(0.264\,65°) = 0.53°$ (to two significant figures).

25.79

The spacing between adjacent maxima in the diffraction grating is given by Eq. (25.9) to be

$$\Delta y \approx \frac{s\lambda}{a} \propto \frac{1}{a},$$

where a is the separation between adjacent slits in the grating. As the number of lines per unit length in the grating doubles successively from 200 lines/cm to 400 lines/cm and 800 lines/cm, a successively decreases to $\frac{1}{2}$ and $\frac{1}{4}$ of the original value, causing $1/a$, and therefore Δy, to increase by a factor of 2 and 4, successively. So the spectra line spacing is doubled and quadrupled, respectively, as we change from a 200 lines/cm grating to a 400 lines/cm and then to an 800 lines/cm one.

25.91

According to Eq. (25.8) the angular location of the m-th order spectrum satisfies $a\theta_m = \lambda$, where λ is the wavelength and a is the spacing between adjacent slits in the grating. When used on the surface of the Earth in air, in which $\lambda = 550\,\text{nm}$, for the first-order spectrum (with $m = 1$) we have $a\sin\theta_1 = \lambda$, so

$$a = \frac{\lambda}{\sin\theta_1},$$

where $\theta_1 = 20.0°$. Similarly, on the Mongoian surface where the wavelength is λ', for the first-order spectrum we have

$$a = \frac{\lambda'}{\sin\theta'_1},$$

where $\theta'_1 = 18.0°$ is the corresponding angular location of the first-order spectrum when the spectrometer is placed in the Mongoian atmosphere. Equate the two expressions for a above to obtain $\lambda/\sin\theta_1 = \lambda'/\sin\theta'_1$, or $\lambda/\lambda' = \sin\theta_1/\sin\theta'_1$. Now, since $\lambda \approx \lambda_0$, the index of refraction of the Mongoian atmosphere is, by definition,

$$n = \frac{\lambda_0}{\lambda'} \approx \frac{\lambda}{\lambda'} = \frac{\sin\theta_1}{\sin\theta'_1} = \frac{\sin 20.0°}{\sin 18.0°} = 1.11\,.$$

25.97

The angular resolution θ_a of a circular aperture of diameter D is given by Eq. (25.15): $\theta_a \approx 1.22\lambda/D$, where λ is the wavelength. In this case the aperture is the objective mirror, with $D = 508\,\text{cm} = 5.08\,\text{m}$, and the wavelength is $550\,\text{nm} = 550 \times 10^{-9}\,\text{m}$. Thus the angular resolution of the mirror is

$$\theta_a \approx \frac{1.22\lambda}{D} = \frac{1.22(550 \times 10^{-9}\,\text{m})}{5.08\,\text{m}} = 1.32 \times 10^{-7}\,\text{rad}\,,$$

which is equivalent to $(1.32 \times 10^{-7}\,\text{rad})(180°/\pi\,\text{rad}) = 7.57 \times 10^{-6}$ degrees. In terms of seconds, this is $(7.57 \times 10^{-6}\,\text{degrees})(60\,\text{min/degree})(60\,\text{s/min}) = 2.72 \times 10^{-2}\,\text{s}$.

26 *Special Relativity*

Answers to Selected Discussion Questions

•26.1

(a) You would see yourself just behind where you are at the instant you spin around, moving backward in space and time until you plop down into the chair. The most distant image would take the longest time to reach you. (b) To see Lincoln, you need only fly away at a speed greater than c, overtake the wavefront corresponding to the desired event, pass it, and turn around. You can rush off at $v \approx 240c$, travel for a month or so, set up a telescope and watch yourself being born (assuming the great event occurred near a window with the shades up). (c) Of course none of this is possible since v must be less than c.

•26.5

No. The scissors' contact point is more a mathematical notion than a physical one. There is no transport of energy from one point in space to another. Nothing actually goes outward from the pivot along the blades. The situation is the same for the overlap region of the two laserbeams. It moves faster than c, but doesn't communicate anything. A long shadow can move faster than c.

•26.7

The speed will remain constant at c, but there will be a Doppler shift lowering the frequency of the light. Since $E = hf$, a drop in frequency corresponds to a reduction in total energy. Because $E = pc$, the momentum must decrease as well. Photons have zero rest-energy and that doesn't change.

•26.9

An electron moving along the wire, say, to the right, sees all the electrons in the first wire to be at rest and all the positive ions to be moving left — the electron sees the other wire move left. The observer electron in wire 2 sees a contraction of wire 1 and a resulting increase of positive charge density while the negative charge distribution is unchanged. The observer electron and all its fellows are attracted to the increased positive charge density — the wires attract.

•26.11

As $V \to c$, $\gamma \to \infty$, and the total energy of the object becomes infinite. We can conclude that it takes an infinite amount of energy to bring a body up to light speed and therefore no object with mass can attain a speed of c.

•26.13

Rewrite Eq. (28.7) as

$$\frac{v_{PO}}{c} = \frac{\dfrac{v_{PO'}}{c} + \dfrac{v_{O'O}}{c}}{1 + \left(\dfrac{v_{PO'}}{c}\right)\left(\dfrac{v_{O'O}}{c}\right)}.$$

If $v_{PO'}/c = 1$ or $v_{O'O}/c = 1$ then $v_{PO}/c = 1$. When $v_{PO'}$ and $v_{O'O}$ are in the same direction, both + or both –, the denominator is > 1 and v_{PO} is less than its classical value, which keeps v_{PO} from exceeding c. When the signs of and are different and the motions are tending to cancel (as when someone runs to the rear of a moving train and stands still with respect to the platform), the denominator is < 1 and v_{PO} is larger than the classical value. That situation allows a light beam moving at $v_{PO'} = +c$ in a system moving at $v_{O'O} = -c$ to still be seen by someone at relative rest to be traveling at c.

•26.15

We generalize the assumption and posit that *an object at one location in the Universe cannot affect another object that is a finite distance away instantaneously.* Then it follows that nothing can pass from one to another without a lapse of time and therefore *nothing can travel infinitely fast.* If there is an upper-limit to speed, and if it is measured to be different in different inertial systems, then it is possible to exceed the speed limit, and that contradicts the premise.

Answers to Odd-Numbered Multiple Choice Questions

1. d **3.** c **5.** b **7.** a **9.** e **11.** c **13.** a
15. d **17.** c

Solutions to Problems

26.9

According to the text, if an event lasts a time duration Δt_S in its own rest frame, then to a moving observer the corresponding time interval is $\Delta t_M = \Delta t_S/\sqrt{1-\beta^2}$ [see Eq. (26.2)], where v is the speed of the moving observer relative to the stationary frame. In this case the event (the joke-telling) lasts a time duration 60.0 s on Earth, in which the event itself is stationary. So by definition $\Delta t_S = 60.0\,\mathrm{s}$. Now, as the observer in the rocket ship flies by at a speed of $v = 0.995c$ he measures the time interval of the same event to be Δt_M, so

$$\Delta t_M = \frac{\Delta t_S}{\sqrt{1-v^2/c^2}} = \frac{60.0\,\mathrm{s}}{\sqrt{1-(0.995c/c)^2}} = 601\,\mathrm{s}.$$

From the moving observer's perspective, the event unfolds in slow motion, with everybody laughing together or not at all.

26.21

The proper length L_S of the platform is related to L_M, its moving length measured by an observer traveling at a speed v relative to the platform, via Eq. (26.5): $L_M = L_S\sqrt{1-v^2/c^2}$. We know that an observer moving at $v = 0.600c$ relative to the platform finds its length to be $L_M = 400.0\,\mathrm{m}$. So the proper length of the platform is

$$L_S = \frac{L_M}{\sqrt{1-v^2/c^2}} = \frac{400.0\,\mathrm{m}}{\sqrt{1-(0.600c/c)^2}} = 500\,\mathrm{m}.$$

26.23

The measured length of 1.000 yd of the paint by an Earth-based observer is its moving length L_M, which is related to the corresponding proper length L_S via Eq. (26.5): $L_M = L_S\sqrt{1-v^2/c^2}$, where v is the speed of the rocket bearing the paint, relative to the Earth. Plug in $L_S = 1.000\,\mathrm{m}$ and $L_M = 1.000\,\mathrm{yd} = (1.000\,\mathrm{yd})(3\,\mathrm{ft/yd})(0.3048\,\mathrm{m/ft}) = 0.9144\,\mathrm{m}$ and solve for v:

$$v = c\sqrt{1-\left(\frac{L_M}{L_S}\right)^2} = c\sqrt{1-\left(\frac{0.9144\,\mathrm{m}}{1.000\,\mathrm{m}}\right)^2} = 0.4048\,c.$$

26.27

The 1.000-m length of the bar is its proper length L_S. As the bar moves by the telescope at a speed v relative to the scope, its moving length L_M measured by the observer through the scope is $L_M = L_S\sqrt{1-v^2/c^2}$. At $v = 0.600c$, the time t it takes for the entire length of the bar to fly past the cross hairs of the telescope is therefore

$$t = \frac{L_M}{v} = \frac{L_S\sqrt{1-v^2/c^2}}{v} = \frac{(1.000\,\text{m})\sqrt{1-(0.600c/c)^2}}{0.600(2.998\times10^8\,\text{m/s})} = 4.45\times10^{-9}\,\text{s} = 4.45\,\text{ns}\,.$$

26.29

Put the x-axis of a coordinate system in the direction of motion of the spaceship (relative to the Earth), and the y-axis is then perpendicular to the direction of motion. If the mast has a proper length L, then (see the diagram to the right)

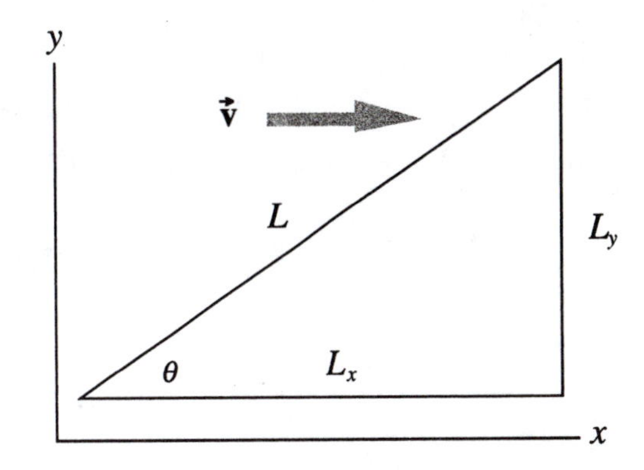

$$\tan\theta = \frac{L_y}{L_x}\,,$$

where $\theta = 21.0°$ is the angle between the orientation of the mast and the direction of motion of the spaceship. Now, when viewed by someone on Earth, which is moving at a speed v relative to the spaceship, L_x will shrink by a factor of $\sqrt{1-v^2/c^2}$ to L'_x because of the Lorentz contraction, while L_y stays the same (since the ship is not moving relative to the Earth in the y-direction). From Eq. (26.5) we have

$$L'_x = L_x\sqrt{1-v^2/c^2} = L_x\sqrt{1-\left(\frac{0.852c}{c}\right)^2} = 0.523\,54\,L_x\,,$$

and so the angle θ now changes to θ', which satisfies

$$\tan\theta' = \frac{L_y}{L'_x} = \frac{L_y}{0.523\,54\,L_x} = 1.910\,07\tan\theta = (1.910\,07)(\tan 21.0°) = 0.733\,2\,.$$

The value of the angle between the mast and the direction of motion of the spaceship, as seen from the Earth, is then $\theta' = \tan^{-1} 0.733\,2 = 36.2°$.

26.39

Denote the Earth, the spaceship, and the laserbeam with subscripts E, S and L, respectively. If the direction from the Earth to the spaceship (in which the laserbeam is projected) is chosen as positive, then v_{LE}, the velocity of the laserbeam relative to the Earth, is $+c$; while v_{SE}, the velocity of the ship relative to the Earth, is $-v$, where the minus sign indicates that the ship is traveling towards the Earth rather than away from it, i.e., in the direction opposite to that of the laserbeam. This gives $v_{ES} = -v_{SE} = +V$. We are looking for v_{LS}, the velocity of the laserbeam (L) relative to the ship (S). To this end we apply Eq. (26.7):

$$v_{LS} = \frac{v_{LE} + v_{ES}}{1 + \frac{v_{LE} v_{ES}}{c^2}} = \frac{c + V}{1 + \frac{cV}{c^2}} = \frac{c(c + V)}{c + V} = c\,,$$

which is indeed independent of the value of V, as required by the Second Postulate of Special Relativity.

26.63

According to Eq. (26.12), the total energy E is the sum of the kinetic energy KE and the rest energy E_0: $E = KE + E_0$. Since the total energy of the muon is $E = 106.7\,\text{MeV}$ while its rest energy is $E_0 = 105.7\,\text{MeV}$, the kinetic energy of the muon must be

$$KE = E - E_0 = 106.7\,\text{MeV} - 105.7\,\text{MeV} = 1.0\,\text{MeV}\,.$$

26.79

The momentum p of the proton of rest mass m_p moving at a speed v is given by Eq. (26.8), $p = \gamma m_p v$, where $\gamma = 1/\sqrt{1 - v^2/c^2}$. Now, we know that the rest energy of the proton is $m_p c^2 = E_0 = 938.272\,\text{MeV}$, so the expression for p may be rewritten as

$$p = \gamma m_p v = \frac{\gamma(m_p c^2)v}{c^2} = \frac{\gamma E_0 v}{c^2} = \frac{E_0 v}{c^2\sqrt{1 - v^2/c^2}}\,,$$

which is an equation for v. To solve for it, first square both sides of the equation to obtain $p^2 = E_0^2 v^2/c^2(c^2 - v^2)$ which, upon rearrangement, becomes $\left[1 + (E_0/pc)^2\right] v^2 = c^2$. Thus

$$v = \frac{c}{\sqrt{1 + \left(\frac{E_0}{pc}\right)^2}} = \frac{c}{\sqrt{1 + \left[\frac{938.272\,\text{MeV}}{(100.0/c)c}\right]^2}} = 0.1060\,c\,.$$

27 The Origin of Modern Physics

Answers to Selected Discussion Questions

•27.1

The NaCl disassociates into ions; namely, Na^+ and Cl^-. The chlorine carries its extra electron to the positive anode, where it gives it up and becomes neutral. The positive sodium ion, deficient by one electron (having given it to the chlorine), migrates to the cathode, where it picks up an electron and becomes neutral. The sodium atoms are not soluble and come out on the cathode as metallic sodium. The net result is that an electron has in effect traveled from cathode to anode that's the current. Had we started with a water solution, the water would have participated in the electrolysis and made things more complicated.

•27.3

If it's finite in size, one might immediately ask, how is the charge distributed (that is, where on the electron is it)? If the electron has no parts, its charge has no parts either, and it becomes difficult to imagine any complicated distribution. Indeed if the charge were, say, spread over the surface of the electron (if it has a surface) what would keep the electron from blowing up under the Coulomb repulsion? Perhaps mass and charge are inseparable in the sense that whatever has mass has charge and vice versa.

•27.7

The atoms in the first block are driven into oscillation in the zy-plane by the E-field, which is transverse to the x-axis. The beam traveling toward the second block is polarized. Atoms in the second block can only oscillate along the z-axis, and radiate in the xy-plane.

•27.11

The wave number $109\,677\,\text{cm}^{-1}$ is the Series Limit of the Lyman Series just as $27\,419\,\text{cm}^{-1}$ is the Series Limit of the Balmer Series. All the levels correspond to the wave numbers of the successive Series Limits. It appears as if the atom, too, must have some sort of internal level structure such that a transition from one to another liberates light of a certain wavelength.

Answers to Odd-Numbered Multiple Choice Questions

1. d **3.** b **5.** b **7.** c **9.** b **11.** a **13.** b
15. c **17.** a

Solutions to Selected Problems

27.7

The total amount of charge passing through the molten bath of NaCl with a current I (= 50.0 A) running for a time duration Δt (= 5.00 min = 300 s) is $q = I\Delta t = (50.0\,\text{A})(300\,\text{s}) = 1.50 \times 10^4\,\text{C}$. Now, to liberate one mole of monovalent ions (such as Na^+ in this case) we need to supply one faraday of charge (1 faraday = 96 485 C/mol). So if we divide the charge q supplied to the NaCl by one faraday, we then get n, the number of moles of sodium liberated:

$$n = \frac{q}{1\,\text{faraday}} = \frac{1.50 \times 10^4\,\text{C}}{96\,485\,\text{C/mol}} = 0.155\,46\,\text{mol}\,.$$

Multiply this by the molar mass of sodium, $m = 22.99\,\text{g/mol}$, to obtain the mass M of the sodium liberated:

$$M = nm = (22.99\,\text{g/mol})(0.155\,46\,\text{mol}) = 3.57\,\text{g}\,.$$

27.13

The molar mass of copper is 63.55 g/mol, so 63.55 g of copper is one gram-mole, which contains one mole, or $N_A = 6.022 \times 10^{23}$, copper atoms. Fro simplicity let's assume that each copper atom can be roughly considered a cube of side length a, and that the spacing between adjacent atoms can be neglected (neither of which is true, of course), then the total volume occupied by this many copper atomic "cubes" would be $N_A a^3$, which would give V, the total volume of the copper block:

$$N_A a^3 = V = \frac{m}{\rho}.$$

Here we noted that V can be found from the mass m ($= 63.55\,\text{g}$) and the density ($\rho = 8.96\,\text{g/cm}^3$) of the copper block to be $V = m/\rho$. Solve for a, the approximate size of the copper atom:

$$a = \left(\frac{m}{\rho N_A}\right)^{1/3} = \left[\frac{63.55\,\text{g}}{(8.96\,\text{g/cm}^3)(6.022 \times 10^{23})}\right]^{1/3} = 2.28 \times 10^{-8}\,\text{cm} = 0.228\,\text{nm}.$$

27.17

To traverse through the crossing E- and B-fields undeflected, the speed v of the electron must satisfy $v = E/B$ (see Chapter 19). Also, the magnetic force exerted on the electron, $F_M = evB$, is responsible for the centripetal force F_C needed for the electron to move in a circular path of radius r:

$$F_M = evB = F_C = \frac{m_e v^2}{r},$$

or $e/m_e = v/Br$. Plug in $v = E/B$ to obtain

$$\frac{e}{m_e} = \frac{v}{Br} = \frac{E/B}{Br} = \frac{E}{B^2 r},$$

which expresses the charge-to-mass ratio of the electron as a function of the E- and B-fields, along with the radius r of the circular path of the electron. In the present problem we are given $r = 15.00\,\text{cm} = 0.150\,0\,\text{m}$ and $E = 20.0\,\text{kV/m} = 20.0 \times 10^3\,\text{V/m}$, so the B-field can be solved from the formula above to be

$$B = \sqrt{\frac{E m_e}{e r}} = \sqrt{\frac{(20.0 \times 10^3\,\text{V/m})(9.109\,39 \times 10^{-31}\,\text{kg})}{(1.602 \times 10^{-19}\,\text{C})(0.150\,0\,\text{m})}} = 8.71 \times 10^{-4}\,\text{T}.$$

27.29

X-ray diffraction by crystals satisfy the Bragg equation, Eq. (27.3): $2d \sin\theta_m = m\lambda$. Here d is the spacing between adjacent atomic planes, θ_m is the angle at which the m-th order reflection of the beam occurs, and λ is the wavelength of the incident beam. The first-order reflection angle θ_1 then satisfies

$$\sin\theta_1 = \left.\frac{m\lambda}{2d}\right|_{m=1} = \frac{\lambda}{2d}.$$

Plug in $d = 0.303\,\text{nm}$ and $\lambda = 0.090\,\text{nm}$, and solve for θ_1:

$$\theta_1 = \sin^{-1}\frac{\lambda}{2d} = \sin^{-1}\left[\frac{0.090\,\text{nm}}{2(0.303\,\text{nm})}\right] = 8.5^\circ\,.$$

27.49

The closest distance R the alpha particle (α) can come to the center of the nucleus is $R = 4kZe^2/m_\alpha v_\alpha^2$ [see Eq. (27.4)]. Here Z is the atomic number of the nucleus, e the elementary charge, m_α the mass of the alpha particle, and v_α its initial speed. Plug in $Z = 26$ (for iron), $m_\alpha = 6.6 \times 10^{-27}\,\text{kg}$, and $v_\alpha = \text{c}/20$ to obtain

$$R = \frac{4kZe^2}{m_\alpha v_\alpha^2} = \frac{4(8.99 \times 10^9\,\text{N}\cdot\text{m}^2/\text{C}^2)(26)(1.602 \times 10^{-19}\,\text{C})^2}{(6.6 \times 10^{-27}\,\text{kg})\left[(2.998 \times 10^8\,\text{m/s})/20\right]^2} = 1.6 \times 10^{-14}\,\text{m}\,.$$

28 *The Evolution of Quantum Theory*

Answers to Selected Discussion Questions

•28.3

Classically, the re-emitted X-rays should come off in a spherical wave and both detectors should have picked up the radiation simultaneously. Since they didn't, the photon picture is upheld.

•28.5

$Nhf = \mathrm{KE}_{\mathrm{max}} + \phi$, just increase the energy imparted to the electron by a factor of N. The maximum KE increases and so the stopping potential must increase, too. $\mathrm{KE}_{\mathrm{max}} = eV_{\mathrm{S}}$. The work function is determined by the metal and does not change unless the metal itself is altered. $Nhf = \phi_0$, and so the threshold frequency $f_0 = \phi/Nh$ must decrease.

•28.7

The light-quantum must provide an energy of at least . Hence, the photon must have a minimum frequency $f_0 = \mathrm{E_a}/h$, very much as in the photoelectric effect.

•28.9

The initial momentum as seen in the center-of-mass system is zero, as is the final momentum. The electron is seen to be at rest after the collision. The initial energy is hf for the photon and $E = \gamma mc^2$ for the moving electron. The final energy is $\mathrm{E}_0 = mc^2$ since the photon is gone and the electron is at rest ($\mathrm{KE} = 0$). Conservation of Energy yields $hf + \gamma mc^2 = mc^2$, which implies $m > \gamma m$.

•28.11

Atoms are pumped up to the 4th level from the ground state, emptying the latter. They immediately drop to the 3rd level, which is metastable and thus unlikely to experience many spontaneous emissions. There is then a population inversion between the 3rd and the 2nd levels and a laser transition occurs from the former down to the latter. Rapid decay from the 2nd to the ground state ensures that the inversion will be maintained, an improvement over the 3-level system.

•28.13

Provided the atom was being illuminated by the proper light, it was absorbing and re- emitting photons almost continuously as it bounced up and down, into and from, the lower-excited state. The key was that the higher level was metastable. An atom can only be in one excited state at a time and so whenever the atom was kicked into the higher state, it could not absorb and re-emit via the lower one. The bright light from the atom blinked off whenever it went into the metastable state, and it blinked back on as soon as the atom dropped out of the metastable state.

Answers to Odd-Numbered Multiple Choice Questions

1. c **3.** a **5.** b **7.** b **9.** c **11.** d **13.** b
15. c **17.** b

Solutions to Problems

28.3

We are looking for λ_p, which is given in Eq. (28.4): $\lambda_p T = 0.002\,898\,\mathrm{m\cdot K}$. Plug in $T = (33+273)\mathrm{K} = 306\,\mathrm{K}$ and solve for λ_p:

$$\lambda_p = \frac{0.002\,898\,\mathrm{m\cdot K}}{306\,\mathrm{K}} = 9.47\times10^{-6}\,\mathrm{m} = 9.47\,\mu\mathrm{m}\,,$$

which is in the infrared (IR) range.

28.23

The energy of each of the ultraviolet photons in question is $hf = hc/\lambda$, where $\lambda = 200\,\mathrm{nm}$ is the wavelength. As a photoelectron absorbs this much energy, a part of the energy goes into overcoming the work function ϕ (= 4.31 eV for zinc — refer to Table 28.2) which is needed to free the electron from its bound state inside the atom, while the rest becomes the kinetic energy of the electron. This leads to Eq. (28.8), $hf = hc/\lambda = \mathrm{KE}_{\mathrm{max}} + \phi$, which we solve for $\mathrm{KE}_{\mathrm{max}}$, the maximum kinetic energy of the emitted electrons:

$$\mathrm{KE}_{\mathrm{max}} = \frac{hc}{\lambda} - \phi = \frac{1240\,\mathrm{eV\cdot nm}}{200\,\mathrm{nm}} - 4.31\,\mathrm{eV} = 1.89\,\mathrm{eV}\,.$$

(Note that, since here λ is given in nm and ϕ in eV, it is more convenient to express hc in terms of eV·nm, rather than the SI units.)

28.25

When the retarding voltage V_{S} reaches such a value that even the photoelectrons which started with the maximum amount of kinetic energy, $\mathrm{KE}_{\mathrm{max}}$, are stopped short of reaching the collector, then the current will decrease to zero. This happens when $\mathrm{KE}_{\mathrm{max}} = eV_{\mathrm{S}}$ [see Eq. (28.6)], so the maximum speed of the photoelectrons satisfies

$$\frac{1}{2} m_{\mathrm{e}} v_{\mathrm{max}}^2 = \mathrm{KE}_{\mathrm{max}} = eV_{\mathrm{S}}\,.$$

In this case the current reaches zero when V_{S} is raised to 1.25 V, so

$$v_{\mathrm{max}} = \sqrt{\frac{2eV_{\mathrm{S}}}{m_{\mathrm{e}}}} = \sqrt{\frac{2(1.602 \times 10^{-19}\,\mathrm{C})(1.25\,\mathrm{V})}{9.109\,39 \times 10^{-31}\,\mathrm{kg}}} = 6.63 \times 10^5\,\mathrm{m/s}\,.$$

Note that, since $v_{\mathrm{max}}/\mathrm{c} = 2.21 \times 10^{-3} \ll 1$, we are justified in using the classical approximation $\mathrm{KE} = \frac{1}{2}mv^2$. Also, as an alternative, you can take $eV_{\mathrm{S}} = 1.25\,\mathrm{eV}$ and $m_{\mathrm{e}} = 0.511\,\mathrm{MeV/c^2}$ to find $v_{\mathrm{max}} = 2.21 \times 10^{-3}\mathrm{c} = 6.63 \times 10^5$ m/s.

28.39

The initial value of the total energy of the electron before its impact with the metal target is $\mathrm{E_i} = \gamma_{\mathrm{i}} m_{\mathrm{e}} \mathrm{c}^2$, where $m_{\mathrm{e}}\mathrm{c}^2$ is the rest energy of the electron and $\gamma_{\mathrm{i}} = 1/\sqrt{1 - v_{\mathrm{i}}^2/\mathrm{c}^2}$, with $v_{\mathrm{i}} = 1.00 \times 10^8$ m/s the initial speed of the electron. After the impact the final value of its total energy becomes $\mathrm{E_f} = \gamma_{\mathrm{f}} m_{\mathrm{e}} \mathrm{c}^2$, where $\gamma_{\mathrm{f}} = 1/\sqrt{1 - v_{\mathrm{f}}^2/\mathrm{c}^2}$, with $v_{\mathrm{f}} = \frac{1}{2} v_{\mathrm{i}} = \frac{1}{2}(1.00 \times 10^8\,\mathrm{m/s}) =$ 0.500×10^8 m/s the final speed of the electron. From the Conservation of Energy we know that the kinetic energy lost by the electron as a result of the collision becomes the energy $\mathrm{E_p}$ of the photon, whose wavelength is λ:

$$\mathrm{KE_i} - \mathrm{KE_f} = (\gamma_{\mathrm{i}} - \gamma_{\mathrm{f}})\, m_{\mathrm{e}} \mathrm{c}^2 = \mathrm{E_p} = \frac{hc}{\lambda}\,.$$

Plug in $\gamma_{\mathrm{i}} = 1/\sqrt{1 - v_{\mathrm{i}}^2/\mathrm{c}^2} = 1/\sqrt{1 - [(1.00 \times 10^8\,\mathrm{m/s})/(2.998 \times 10^8\,\mathrm{m/s})]^2} = 1.060\,749$ and $\gamma_{\mathrm{f}} = 1/\sqrt{1 - v_{\mathrm{f}}^2/\mathrm{c}^2} = 1/\sqrt{1 - [(0.500 \times 10^8\,\mathrm{m/s})/(2.998 \times 10^8\,\mathrm{m/s})]^2} = 1.014\,204$, along with $m_e c^2 = 0.510\,999\,\mathrm{MeV}$ and $hc = 1240\,\mathrm{eV\cdot nm}$, and solve for λ:

$$\begin{aligned}\lambda &= \frac{hc}{(\gamma_i - \gamma_f)\, m_e c^2} \\ &= \frac{1240\,\mathrm{eV\cdot nm}}{(1.060\,749 - 1.014\,204)(0.510\,999 \times 10^6\,\mathrm{eV})} \\ &= 5.22 \times 10^{-11}\,\mathrm{m} = 52.2\,\mathrm{pm}\,.\end{aligned}$$

Note that, since $v_i/c \approx 1/3$, v_i is not quite negligible compared with c; so relativistic approach must be followed to ensure sufficient accuracy for λ (to three significant figures). In fact if we were to use the classical approximation $\mathrm{KE} = \frac{1}{2}mv^2$ instead, then the answer obtained for λ would be 58.2 pm, which amounts to a 10% difference from the (more accurate) relativistic value obtained above.

28.47

As a hydrogen atom makes a transition from an initial state (i) to a final state (f), it emits a photon whose energy E_p is the difference between the initial and final energies of the atom (due to the Conservation of Energy). In our case the initial state corresponds to $n = 3$ while the final one is $n = 2$. Thus

$$E_p = E_i - E_f = E_3 - E_2 = \left(-\frac{13.6\,\mathrm{eV}}{3^2}\right) - \left(-\frac{13.6\,\mathrm{eV}}{2^2}\right) = 1.89\,\mathrm{eV}\,,$$

where we noted that $E_n = -(13.6\,\mathrm{eV})/n^2$. Then from $E_p = hf = hc/\lambda$ we find the corresponding frequency f and wavelength λ of the photon to be

$$f = \frac{E_p}{h} = \frac{1.89\,\mathrm{eV}}{4.136 \times 10^{-15}\,\mathrm{eV\cdot s}} = 4.57 \times 10^{14}\,\mathrm{Hz}$$

and

$$\lambda = \frac{hc}{E_p} = \frac{1240\,\mathrm{eV\cdot nm}}{1.89\,\mathrm{eV}} = 656\,\mathrm{nm}\,.$$

(Note that λ can also be obtained from $\lambda = c/f$.)

28.63

The electron in the n-th Bohr orbit of a hydrogen-like atom revolves about the atomic nucleus with a speed v_n in a circular orbit of radius r_n, making one complete revolution in a time interval $T_n = 2\pi r_n/v_n$ (note that in each revolution the electron covers a distance of $2\pi r_n$, the circumference of its circular orbit). The corresponding frequency f_n is then

$$f_n = \frac{1}{T_n} = \frac{v_n}{2\pi r_n}\,.$$

To find f_n we need to know the expressions for v_n and r_n. The orbital radius r_n is given by Eq. (28.17): $r_n = n^2\hbar^2/m_e k_0 Ze^2$, while the orbital speed v_n can be found in terms of r_n from Eq. (28.15), the quantization condition for the orbital angular momentum: $m_e v_n r_n = n\hbar$. Substitute both expressions into the formula for f_n above to find

$$f_n = \frac{v_n}{2\pi r_n} = \frac{n\hbar/m_e r_n}{2\pi r_n} = \frac{n\hbar}{2\pi m_e r_n^2} = \frac{n\hbar}{2\pi m_e \left(\dfrac{n^2\hbar^2}{m_e k_0 Ze^2}\right)^2} = \frac{m_e k_0^2 Z^2 e^4}{2\pi n^3 \hbar^3}.$$

29 Quantum Mechanics

Answers to Selected Discussion Questions

•29.1

Since $\lambda = 1.25/\sqrt{V}\,\text{nm}$, the wavelength can be quite small. For example, at 100 kV, $\lambda = 0.004\,\text{nm}$. Although electrons can be focused easily, X-rays of comparable wavelength are very difficult to focus (though progress is being made in that endeavor). Electrons (in vacuum to prevent scattering) pass through a thin slice of the sample where they are scattered off in a pattern that corresponds to the information. This beam is collected by the objective, which forms a magnified intermediate image. The electrons from that image are collected and the image further enlarged by the projection lens. The product of these two magnifications can be an enlargement of 200 000× or so.

•29.5

According to classical theory, a particle passes through one hole, a wave through both holes. If either coil records the presence of an electron, we are dealing with the particle-like manifestation of the electron and the wavelike behavior will vanish. Thus, the theory maintains that the interference pattern will vanish as soon as either coil picks up the transit of an electron. Presumably, the induced current will produce an induced magnetic field, which will destroy the interference.

•29.7

If the particle is confined, there will be a nonzero Δx and therefore a nonzero Δp, which is the minimum p the particle must have and it corresponds to a minimum KE $= p^2/2m$. The atoms in a box at absolute zero must themselves still be moving around with a zero-point energy.

•29.9

A range of frequencies and wavelengths is needed to synthesize the packet and, hence, a range of frequency (energy) and wavelength (momentum) is always present, thus constituting an uncertainty in E and p for the particle. To make $\Delta p = 0$, we must use a monochromatic wave, one wavelength, one p, and no uncertainty. But where will the particle be if the wave is a perfect sine wave? Anywhere, and so $\Delta x = \infty$. If the packet shrinks so $\Delta x \to 0$, the number of sine waves needed to make the packet increases to infinity, and $\Delta p \to \infty$.

•29.11

The peaks occur at the Alkali Metals, which have a single outer electron in an unfilled shell. This electron is shielded from the nucleus by all the filled shells below it, and so is held weakly. If the nuclear charge is $+Ze$, the electron "sees" a charge of just +e. Accordingly, the electron cloud is relatively large. Similarly, boron, aluminum, and gallium atoms have three outer electrons that see a nuclear charge of $+3e$ and are strongly bound in small orbits. Thus, the curve rises and falls.

Answers to Odd-Numbered Multiple Choice Questions

1. b **3.** d **5.** a **7.** d **9.** d **11.** c **13.** d
15. a

Solutions to Problems

29.3

The speed of the electron, $v = c/10$, corresponds to a γ- value of $\gamma = 1/\sqrt{1 - v^2/c^2} = 1/\sqrt{1 - (1/10)^2} = 1.005 \approx 1$, and so classical approximation can be used to yield an accuracy of over 99% in both p and λ. We may therefore substitute the classical expression $m_e v$ for p, the momentum of the electron, into the de Broglie formula $\lambda = h/p$ to find the wavelength λ of the electron of mass m_e moving at the speed v:

$$\lambda = \frac{h}{p} \approx \frac{h}{m_e v} = \frac{6.626 \times 10^{-34}\,\text{J}\cdot\text{s}}{(9.109\,39 \times 10^{-31}\,\text{kg})\,[(2.998 \times 10^8\,\text{m/s})/10]} = 2.4 \times 10^{-11}\,\text{m}\,,$$

or 24 pm. (You may compare this result with that of the relativistic formula, $\lambda = h/\gamma m_e v$. It turns out that the two values for λ are identical up to two significant figures.)

29.7

Suppose that the electron is accelerated by a voltage difference of V. Then as it emerges from the accelerating potential the electron has acquired a kinetic energy in the amount of $\text{KE} = eV$, and so the momentum p of the electron is given by $p^2/2m_e = \text{KE} = eV$ to be $p = \sqrt{2eVm_e}$. The corresponding de Broglie wavelength of the electron follows from Eq. (29.1) to be

$$\lambda = \frac{h}{p} = \frac{h}{\sqrt{2eVm_e}}\,.$$

Plug in $\lambda = 0.10\,\text{nm} = 0.10 \times 10^{-9}\,\text{m}$ and solve for V, the required accelerating voltage:

$$V = \frac{h^2}{2em_e\lambda^2} = \frac{(hc)^2}{2e(m_e c)^2\lambda^2} = \frac{(1240\,\text{eV}\cdot\text{nm})^2}{2e(0.511 \times 10^6\,\text{eV})(0.10\,\text{nm})^2} = 0.15\,\text{kV}\,.$$

Note that we multiplied both the denominator and the numerator by c^2 so as to use the well-known data of $hc = 1240\,\text{eV}\cdot\text{nm}$ and $m_e c^2 = 0.511\,\text{MeV}$. This is numerically a little simpler to handle than converting everything to SI units. Also note that by definition $1\,\text{eV}/1\,\text{V} = 1e$. Finally, since $\text{KE} = eV = 0.15\,\text{keV} \ll m_e c^2 = 0.511\,\text{MeV}$, we are justified in using non-relativistic approach.

29.27

The K-shell can accommodate 2 electrons, L can have 8, M can have 18, and N can have as many as 32. To see this, first we note that each distinctive quantum state specified by a complete set of quantum numbers, namely $(n,\, l\,, m_l,\, m_s)$, can only be occupied by a single electron — as required by the Pauli Exclusion Principle. So the number of electrons each shell can accommodate is simply the number of distinct quantum states that are available in the given shell (with a fixed value of n).

For example, for the K-shell we have $n = 1$, and since l can only assume integer values between 0 and $n-1$, its only allowed value is 0, as does m_l, which can go from $-l$ to $+l$. This leaves only m_s to be variable between $+\frac{1}{2}$ (spin-up) and $-\frac{1}{2}$ (spin-down), so the only two distinct quantum states for the K-shell are $(1,\,0,\,0,\,+\frac{1}{2})$ and $(1,\,0,\,0,\,-\frac{1}{2})$ — and the number of electrons that can be accommodated by that shell is 2.

In general, for a given shell (of fixed n- value), since the number of different m_l values for a given number l is $2l+1$ (as m_l goes from $-l$ to $+l$), while l itself goes from $l_{\min}=0$ up to $l_{\max}=n-1$, the number of distinct combinations of l and m_l is $(2l_{\min}+1)+\cdots+(2l_{\max}+1)=1+3+5+\cdots+(2n-1)$. Multiplying this by 2, which accounts for the two choices available for m_s, we obtain the general formula for N_n, the number of electrons that can be accommodated in a state of given n:

$$N_n = 2[1+3+5+\cdots+(2n-1)]\,,$$

which can be summed up to yield $2n^2$. You may check this by substituting $n=1$ (for the K-shell): $N_2=2(1)^2=2$; $n=2$ (for the L-shell): $N_2=2(2)^2=8$; etc.

29.43

We are given the mean lifetime of the rho meson to be $T=4.4\times10^{-24}$ s. Thus the uncertainty in T is $\Delta T\sim T=4.4\times10^{-24}$ s. It follows from the Uncertainty Principle, $\Delta\mathrm{E}\,\Delta T\geq\frac{1}{2}\hbar$, or $\Delta\mathrm{E}\geq\hbar/2\Delta T$, that the minimum uncertainty in energy for the meson is

$$\Delta\mathrm{E}(\min)=\frac{\hbar}{2\Delta T}\sim\frac{6.582\,12\times10^{-16}\,\mathrm{eV\cdot s}}{2(4.4\times10^{-24}\,\mathrm{s})}=7.5\times10^{7}\,\mathrm{eV}=75\,\mathrm{MeV}\,,$$

which we divide by $\mathrm{E}=765$ MeV, the rest energy of the meson, to find the fractional uncertainty in E to be

$$\frac{\Delta\mathrm{E}}{\mathrm{E}}\sim\frac{75\,\mathrm{MeV}}{765\,\mathrm{MeV}}=0.98=9.8\%\,.$$

29.47

The uncertainty in position, Δx, for a particle of mass m moving at a speed v_x in the x-direction is related to the uncertainty in the x-component of its momentum via the Uncertainty Principle, Eq. (29.5): $\Delta p_x\Delta x\geq\frac{1}{2}\hbar$, which can be rewritten, with $p_x=mv_x$, as

$$\Delta p_x\Delta x=\Delta(mv_x)\Delta x=m\Delta v_x\Delta x\geq\frac{1}{2}\hbar.$$

Thus $\Delta x(\min)=\hbar/2m\Delta v_x$. In this case we are given that $\Delta p_x/p_x=\Delta(mv_x)/mv_x=m\Delta v_x/mv_x=\Delta v_x/v_x=1/1000$, $m=10.0\,\mathrm{g}=10.0\times10^{-3}\,\mathrm{kg}$, and $v_x=20.0\,\mathrm{cm/s}=0.200\,\mathrm{m/s}$; and so

$$\begin{aligned}\Delta x(\min)&=\frac{\hbar}{2m\Delta v_x}=\frac{\hbar}{2mv_x(\Delta v_x/v_x)}\\&=\frac{1.054\,57\times10^{-34}\,\mathrm{J\cdot s}}{2(10.0\times10^{-3}\,\mathrm{kg})(0.200\,\mathrm{m/s})(1/1000)}\\&=2.64\times10^{-29}\,\mathrm{m}\,.\end{aligned}$$

30 Nuclear Physics

Answers to Selected Discussion Questions

•30.3

The atom-bomb trigger generates temperatures in excess of 100 million K. The lithium is blasted by neutrons and converted into tritium, which then combines via fusion with deuterium, liberating large amounts of energy. The fast neutrons ($\approx 14.1\,\text{MeV}$) then fission the surrounding U-238 "blanket" which can be as large as can be delivered since there's no concern about critical mass. Tritium can be separated from sea water, but it's usually produced by putting lithium-6 in a fission reactor.

•30.5

The surface of the pellet is vaporized, driving the remainder violently inward. The inner core compresses to tremendous densities (1000 times that of water). Rising to temperatures in excess of 100 million degrees, the thermonuclear processes begin and there is a mini-H-bomb explosion (about the equivalent of 50 lb of high explosives). A constant drop and blast is the ultimate goal. ${}^3_1\text{H} + {}^2_1\text{H} \longrightarrow {}^4_2\text{He} + \text{n} + 17.6\,\text{MeV}$.

•30.7

The half-life is the time it takes the sample to decay to half its original value. The nuclear mean lifetime is $1/0.693 = 1.44$ times longer than the half-life — the average radionuclide will live that long. The half-life of an American human is 68 years; half the people born 68 years before will be dead. But the mean human life is certainly not $1.44(68\,\text{y}) = 98\,\text{y}$. The nuclei don't age, we do. Our chances of surviving two half-lives (136 y) are zero. By comparison, some few nuclei will go on for 100 half-lives and more.

Answers to Odd-Numbered Multiple Choice Questions

1. d **3.** b **5.** b **7.** d **9.** b **11.** a **13.** d
15. b **17.** a **19.** e

Solutions to Selected Problems

30.1

In the symbol ${}^{A}_{Z}\mathrm{X}$ which we use for the element X, A represents the number of nucleons in the element while Z is the atomic number (the number of protons) of the element. So in the case of ${}^{111}_{50}\mathrm{Sn}$ the number of nucleons is 111.

30.11

Use the conversion factor between the unified atomic mass (u) and kg: $1\,\mathrm{u} = 1.660\,540 \times 10^{-27}\,\mathrm{kg}$, to convert the mass m of Nb-93 from kg to u:

$$m = \frac{1.542\,748 \times 10^{-25}\,\mathrm{kg}}{1.660\,540 \times 10^{-27}\,\mathrm{kg/u}} = 92.906\,40\,\mathrm{u}\,.$$

While the value of m calculated above is that of the Nb-93 nuclide only (not any of its isotopes), the atomic mass listed for niobium ($92.906\,4\,\mathrm{u}$) is the average value of the atomic masses of the various isotopes of niobium, weighed according to their relative abundances.

30.13

Suppose that bromine has only two stable isotopes, which are listed in the problem statement, namely Br-79, with atomic mass $m_{79} = 78.918\,336\,\mathrm{u}$; and Br-81, with atomic mass $m_{81} = 80.916\,289\,\mathrm{u}$. Then, on the one hand, since the relative abundance of Br-79 is $\rho_{79} = 50.7\%$, the relative abundance for Br-81 must be $X\% = 1 - 50.7\% = 49.3\%$; while on the other hand, $X\%$ can be found from the weighed-average value of the atomic mass of bromine:

$$m = \rho_{79} m_{79} + \rho_{81} m_{81} = (50.7\%)(78.918\,336\,\mathrm{u}) + X\%\,(80.916\,289\,\mathrm{u}) = 79.909\,\mathrm{u}\,,$$

which gives $X\% = 49.3\%$, in agreement with the assumption that only two stable isotopes of bromine be present. So it is unlikely for bromine to have any other long-lived isotopes.

30.21

The radius R of a nucleus is proportional to the (1/3)-th power of its atomic mass number, A, according to Eq. (30.1): $R = R_0 A^{1/3}$. Plug in $R = \frac{1}{2}(7.2\,\text{fm}) = 3.6\,\text{fm}$ and $R_0 \approx 1.2\,\text{fm}$, and solve for A: $A^{1/3} = R/R_0$,

$$A = \left(\frac{R}{R_0}\right)^3 = \left(\frac{3.6\,\text{fm}}{1.2\,\text{fm}}\right)^3 = 3.0^3 = 27\,.$$

(Checking the Periodic Table, we see that the element in question is probably aluminum, ${}^{27}_{13}\text{Al}$).

30.37

An Iron-54 atom is made of 26 protons, 26 electrons, and $54-26 = 28$ neutrons. Before forming the atom, these have a combined mass of $m = 26m_p + 26m_e + 28m_n = 26(m_p + m_e) + 28m_n = 26m_H + 28m_n$, where $m_H = m_p + m_e$ is the mass of a hydrogen atom (which consists of a proton and an electron). The difference between m and the actual mass m_{Fe} of the Iron-54 nuclide is then its mass defect:

$$\begin{aligned}\Delta m &= m - m_{Fe} = (26m_H + 28m_n) - m_{Fe}\\ &= 26(1.007\,825\,\text{u}) + 28(1.008\,665\,\text{u}) - 53.939\,613\,\text{u}\\ &= 0.506\,457\,\text{u}\,.\end{aligned}$$

The total binding energy E_B of the nuclide is given by $E_B = \Delta m\, c^2$, which we divide by A $(= 54)$, the number of nucleons in the nuclide, to obtain the binding-energy-per-nucleon:

$$\frac{E_B}{A} = \frac{\Delta m\, c^2}{A} = \frac{(0.506\,457\,\text{u})(931.494\,\text{MeV/u})}{54} = 8.736\,\text{MeV}$$

(to four significant figures). Note the conversion factor we used: $1\,\text{u} = 931.494\,\text{MeV}/c^2$.

30.41

The decay reaction is ${}^{232}_{92}\text{U} \longrightarrow {}^{228}_{90}\text{Th} + {}^{4}_{2}\alpha$. Before the decay reaction the mass of the ${}^{232}_{92}\text{U}$ *nuclide* is the mass of the ${}^{232}_{92}\text{U}$ *atom* minus that of its 92 electrons: $m_{232} = 232.037\,13\,\text{u} - 92m_e$, where m_e is the mass of an electron. After the decay the total rest mass of the products (an alpha particle plus a ${}^{228}_{90}\text{Th}$ nuclide) is $m_\alpha + m_{228}$, where m_α is the mass of an alpha particle (i.e., a ${}^{4}_{2}\text{He}$ nucleus), which equals the mass of a ${}^{4}_{2}\text{He}$ *atom* minus that of its two electrons: $m_\alpha = m_{He} - 2m_e = 4.002\,603\,\text{u} - 2m_e$; and m_{228} is that of the ${}^{228}_{90}\text{Th}$ nucleus, or that of the ${}^{228}_{90}\text{Th}$ *atom* minus the mass of its 90 electrons: $m_{228} = 228.028\,73\,\text{u} - 90m_e$. So as a result of the decay process the rest mass of the system changes by

$$\begin{aligned}\Delta m &= m_{232} - (m_\alpha + m_{228})\\ &= (232.037\,13\,\text{u} - 92m_e) - [(4.002\,603\,\text{u} - 2m_e) + (228.028\,73\,\text{u} - 90m_e)]\\ &= 0.005\,797\,\text{u}\,.\end{aligned}$$

Note that the electron masses cancel out. This mass difference corresponds to a disintegration energy of

$$Q = \Delta m\, c^2 = (0.005\,797\,\mathrm{u})(931.494\,\mathrm{MeV/u}) = 5.40\,\mathrm{MeV}\,,$$

which is the maximum amount of kinetic energy available to the decay products.

30.47

The difference in mass before and after the reaction $\mathrm{p} + {}^7_3\mathrm{Li} \longrightarrow \alpha + \alpha$ is

$$\begin{aligned}\Delta m &= (m_{\mathrm{p}} + m_{\mathrm{Li}}) - (m_\alpha + m_\alpha)\\ &= 1.007\,825\,\mathrm{u} + 7.016\,003\,\mathrm{u} - 2(4.002\,603\,\mathrm{u})\\ &= 0.018\,622\,\mathrm{u}\,.\end{aligned}$$

Note that here the masses we used were those of the *atoms*, as provided in the problem statement, rather than those of the *nuclides* as we should have used. This does not lead to any error in Δm, however, as the total number of electrons in the ${}^1\mathrm{H}$ and ${}^7_3\mathrm{Li}$ atoms is 4, which is the same as that in two ${}^4_2\mathrm{He}$ atoms. So we have overestimated the mass of the the proton and the ${}^7_3\mathrm{Li}$ nuclide by $4m_{\mathrm{e}}$ while at the same time done the same thing for the two resultant alpha particles — so there is no net error in Δm stemming from using the data for the whole atoms instead of the corresponding nuclides.

The excess kinetic energy carried by the two alpha particles is the energy released as a result of the reduction in the total rest mass of the particles:

$$\mathrm{KE} = \Delta m\, c^2 = (0.018\,622\,\mathrm{u})(931.494\,\mathrm{MeV/u}) = 17.346\,\mathrm{MeV}\,.$$

30.49

Similar to Problem (30.47), the total number of electrons in the ${}^9_4\mathrm{Be}$ and ${}^4_2\mathrm{He}$ atoms is 6, the same as that in the ${}^{12}_6\mathrm{C}$ atom. So when considering the net change in mass in the nuclear reaction ${}^9_4\mathrm{Be} + {}^4_2\alpha \longrightarrow {}^{12}_6\mathrm{C} + {}^1_0\mathrm{n} + Q$ we may use the *atomic* masses of ${}^9_4\mathrm{Be}$, ${}^4_2\mathrm{He}$ and ${}^{12}_6\mathrm{C}$ instead of their respective *nuclear* masses — as the electronic masses before and after the reaction cancel out from the expression — please also see Problem (30.41), in which m_{e} is explicitly shown to drop out from the expression for the mass difference Δm. The value of Q in excess of the incoming kinetic energy of the alpha particle is the energy associated with the net change in mass in the reaction:

$$\begin{aligned}Q &= \Delta m\, c^2 = (m_{\mathrm{Be}} + m_{\mathrm{He}} - m_{\mathrm{C}} - m_{\mathrm{n}})c^2\\ &= (9.012\,182\,\mathrm{u} + 4.002\,603\,\mathrm{u} - 12.000\,000\,\mathrm{u} - 1.008\,665\,\mathrm{u})(931.494\,\mathrm{MeV/u})\\ &= 5.700\,74\,\mathrm{MeV}\,.\end{aligned}$$

30.67

The activity R of a radioactive sample is given by Eq. (30.13), $R = \lambda N$, where λ is the decay constant of the sample and N is the number of radioactive nuclei in the sample. In this case we are concerned with radon-222, for which $\lambda = 2.1 \times 10^{-6}$ decays/s. Also, since the molar mass of radon-222 is about $m_0 = 222\,\text{g/mol}$ the number of radon-222 nuclei in a sample of mass $m = 1.00\,\text{mg}$ is given by $N = (m/m_0)N_A$, with N_A the Avogadro's Number. Plug these expressions into $R = \lambda N$ to obtain

$$\begin{aligned} R = \lambda N &= \frac{\lambda m N_A}{m_0} \\ &= \frac{(2.1 \times 10^{-6}\,\text{decays/s})(1.00 \times 10^{-3}\,\text{g})(6.022 \times 10^{23}/\text{mol})}{222\,\text{g/mol}} \\ &= 5.7 \times 10^{12}\,\text{Bq}\,. \end{aligned}$$

(Note that the unit of R here is decays/s, which is equivalent to Bq.)

31 High-Energy Physics

Answers to Selected Discussion Questions

•31.1

The collider smashes a beam of electrons head-on into a beam of positrons with a combined energy of approximately 100 GeV. This quantity is more than enough to create Z^0 bosons — this is a so-called Z^0 factory, albeit a rather feeble one. The two electron pulses are accelerated (up to 1 GeV), condensed, and injected into the linac (linear accelerator). They are joined by a previously processed positron bunch, which is further accelerated, along with the leading electron bunch, to the end of the linac. The trailing electron bunch is diverted to the side, where it produces a shower of positrons that is sent back to the start for processing so that it can join the next pulse of electrons. Meanwhile, the two high-energy bunches are turned around by magnets at the end of the linac and they collide in the detector.

The drift-tube linac is a succession of tubular conductors that have voltages applied to them just at the right moment so the electrons (or positrons) are accelerated across each gap. The particles see no E-field inside the conductors and drift along them. As they move faster, the cylinders are longer, so the generator frequency can be constant.

•31.3

For the most part, the basis for the distinction between matter and antimatter arises by comparison to the stuff of our ordinary existence. If a baryon (for example, Λ) decays to ordinary matter (Λ decays into a neutron), we call it matter. If it decays to antimatter ($\overline{\Lambda}$ decays into an antineutron), we call it antimatter. But the mesons are half-and-half (quark-antiquark), so we can expect ambiguity. There is no way to determine, and no need to determine, which pion is the antipion.

•31.7

A negatively charged pion entered the chamber leaving a clear slightly curved track, which means that it had a good deal of linear momentum ($R \propto p$). It struck a proton and created two nonionizing neutral particles (K^0 and Λ^0) that left no tracks. Being unstable, they decayed via the weak force (thus living long enough to traverse a substantial distance). The entire process was discussed in Discussion Questions (31.5) and (31.6).

Answers to Odd-Numbered Multiple Choice Questions

1. d **3.** e **5.** c **7.** c **9.** a **11.** a **13.** d
15. a **17.** c **19.** b

Solutions to Problems

31.7

The reaction $\mu^- \longrightarrow e^- + \nu_e + \nu_\mu$ is impossible. In fact, the left side has only a muon so the electron-lepton number is $L_e = 0$; while the right side has an electron (e^-) plus an electron-neutrino (ν_e), for a total electron-lepton number of $L_e = 1 + 1 = 2$. So the conservation of L_e is violated.

31.11

No conservation law is violated in the reaction $K^- \longrightarrow \mu^- + \overline{\nu}_\mu$. It conserves electric charge ($Q_i = -1 = -1 + 0 = Q_f$), spin angular momentum (K^- is spinless; while μ^- and the $\overline{\nu}_\mu$, both spin-$\frac{1}{2}$ particles, may combine their spin angular momenta to yield a net spin of zero), as well as baryon number ($B_i = B_f = 0$, as there are no baryons involved). Also, there is no meson number to worry about. Similar to the case in Problem (31.9), the decay is controlled by the weak force, due to the fact that strangeness is not conserved, K^- being the only particle involved with strangeness. (Note that strangeness must be conserved in the strong interaction.)

31.15

Since the Λ^0 particle is basically at rest before the decay, it has only the rest energy of $m_\Lambda c^2 =$ 1115.6 MeV. After the decay the total rest energy of the particles generated, a π^- and a proton, is $m_\pi c^2 + m_p c^2 = 139.6\,\text{MeV} + 938.3\,\text{MeV} = 1077.9\,\text{MeV}$. The amount of the initial energy in excess of the final rest energy then becomes the kinetic energy of the resulting particles:

$$\text{KE} = m_\Lambda c^2 - (m_\pi c^2 + m_p c^2) = 1115.6\,\text{MeV} - 1077.9\,\text{MeV} = 37.7\,\text{MeV}\,.$$

The decay process is driven by the weak interaction so it does not have to observe the conservation of strangeness. Note that the strangeness changes from $S_i = -1$ for Λ^0 to $S_f = 0$ for the resultant particles.

31.27

The Q value is the energy released in the reaction, excluding the initial kinetic energy of the particles before the reaction. So $Q = \Delta m\, c^2$, where Δm is the difference in rest mass before and after the reaction. In this case we have a π^- plus a proton before the reaction, and a K^0 plus a Λ^0 afterwards. Thus $\Delta m = m_i - m_f = (m_\pi + m_p) - (m_K + m_\Lambda)$, and so

$$\begin{aligned} Q = \Delta m\, c^2 &= (m_\pi + m_p - m_K - m_\Lambda)c^2 \\ &= 139.6\,\text{MeV} + 938.3\,\text{MeV} - 497.7\,\text{MeV} - 1115.6\,\text{MeV} \\ &= -535.4\,\text{MeV} < 0\,, \end{aligned}$$

which means that the rest energy of the final product is higher than that of the initial particles by 535.4 MeV. So this reaction cannot take place spontaneously with the pion and the proton moving slowly towards each other. Rather, the two particles must possess a considerable amount of combined initial kinetic energy (at least 535.4 MeV) to make up for the increase in rest energy as a result of the reaction.

31.31

As a meson the D^0 is made of a quark-antiquark pair, or $q\overline{q'}$. Since $C = +1$ one of the quarks in its configuration must be a c-quark, with $Q = +2/3$, $S = 0$, and $C = +1$. Now, for the meson we have $Q = 0$, $C = +1$, and $S = 0$; so for the next quark we require that $Q = -2/3$, $S = 0$, and $C = 0$, and that makes it a $\overline{u}$. So finally

$$D^0 = c\,\overline{u}\,.$$

31.35

In terms of their quark configurations, the three particles involved in the reaction $K^0 \longrightarrow \pi^+ + \pi^-$ can be expressed as $K^0 = d\overline{s}$, $\pi^+ = u\overline{d}$, and $\pi^- = d\overline{u}$; and so the reaction is equivalent to

$$d\overline{s} \longrightarrow u\overline{d} + d\overline{u}\,.$$

We see that, in addition to a common d-quark present on both sides, the left side has an extra $\overline{s}$-quark while the right side has an extra $\overline{d}$-quark plus a u-$\overline{u}$ pair. What happens is that the $\overline{s}$-quark decays into a $\overline{d}$-quark, liberating energy which is in turn responsible for the creation of the u-$\bar{\text{u}}$ pair. The decay product then consists of the original d-quark, a $\bar{\text{d}}$-quark, plus a u-$\bar{\text{u}}$ pair which breaks up, forming a π^+ $(= \text{u}\bar{\text{d}})$ and a π^- $(= \text{d}\bar{\text{u}})$.

31.39

In the text we obtained from the Uncertainty Principle the Yukawa formula for the range R of a force mediated by a particle of mass m: $R \approx h/4\pi mc$, which gives $m \approx h/4\pi Rc = hc/4\pi Rc^2$. Since we want the final result for m to be in GeV/c^2 while the range R is given in meters, let's convert the unit of hc to from eV·nm to GeV·m:

$$hc = (1240\,\text{eV}\cdot\text{nm})(10^{-9}\,\text{GeV/eV})(10^{-9}\,\text{m/nm}) = 1.240 \times 10^{-15}\,\text{GeV}\cdot\text{m}\,.$$

Plug in this value, along with $R \approx 10^{-31}$ m, to find the approximate mass of the gauge boson:

$$m \approx \frac{hc}{4\pi Rc^2} = \frac{1.240 \times 10^{-15}\,\text{GeV}\cdot\text{m}}{4\pi(10^{-31}\,\text{m})c^2} \approx 10^{15}\,\text{GeV/c}^2\,.$$